"This book brilliantly brings together a series of case examples of people who have the lived experience of being neurodiverse. By using their own narratives it brings to life the issues around neurodiversity in a very thoughtful and creative manner. An additional strength is that it uniquely combines people who would traditionally be described as living with neurodevelopmental, neurological and mental health conditions. Beautifully written; everyone with an interest in this topic should definitely read this book."

Robin Morris, *Professor of Neuropsychology, King's College Institute of Psychiatry, Psychology and Neuroscience*

Redefining Neurodiversity

Redefining Neurodiversity is a transformative exploration of neurodiversity, redefining how we perceive and engage with cognitive differences. By blending scientific research with personal narratives, this book highlights the lived experiences of neurodivergent individuals – challenging outdated deficit-based models and advocating for inclusive, strength-based approaches.

This book brings together voices from a diverse range of neurodivergent individuals, researchers, clinicians, and educators to redefine neurodiversity beyond traditional medical models. It examines neurodevelopmental conditions, acquired neurological differences, and the intersection of mental health and cognitive diversity. Through compelling personal stories, *Redefining Neurodiversity* highlights the real-world challenges and triumphs of neurodivergent individuals, illustrating how systemic barriers in education, employment, and healthcare can be dismantled. The book provides practical recommendations for fostering inclusive environments where neurodivergent people are not merely accommodated but truly valued. It is a call to shift perspectives – from remediation to empowerment and from exclusion to belonging. The book also has been written in support of The Barbara Wilson Centre CIC, a new organisation that provides services to support people who have neurodiverse needs.

This book is an essential read for academics, clinicians, educators, policymakers, industry professionals, and anyone seeking to better understand neurodivergence. Whether you are neurodivergent yourself, work alongside neurodivergent individuals, or wish to build a more inclusive society, *Redefining Neurodiversity* offers insights that are both intellectually rigorous and deeply human.

Sara Simblett is a clinical academic psychologist specialising in neuropsychology, passionate about supporting people with cognitive and emotional differences to live well.

Ashley Polhemus is a trained biomedical engineer, epidemiologist, and product development professional, with a research background.

Rebekah Jamieson-Craig is a clinical psychologist who has an interest in neurological and broader heath conditions and how these impact on the individual.

Faith Matcham is a health psychologist and Associate Professor of Psychology, based at the University of Sussex, Brighton.

Redefining Neurodiversity

Designing Practical Solutions through
Lived Experience

Edited by
Sara Simblett, Ashley Polhemus, Rebekah
Jamieson-Craig and Faith Matcham

R Routledge
Taylor & Francis Group

LONDON AND NEW YORK

Designed cover image: getty images via Eoneren

First published 2026
by Routledge
4 Park Square, Milton Park, Abingdon, Oxon OX14 4RN

and by Routledge
605 Third Avenue, New York, NY 10158

Routledge is an imprint of the Taylor & Francis Group, an informa business

British Library Cataloguing in Publication Data
A catalogue record for this book is available from the British Library

Library of Congress Cataloging-in-Publication Data
A catalog record has been requested for this book

ISBN: 9781032714738 (hbk)
ISBN: 9781032717289 (pbk)
ISBN: 9781032714745 (ebk)

DOI: 10.4324/9781032714745

Typeset in Optima
by Taylor & Francis Books

Contents

Illustrations

Figure

Table

List of Contributors

Lisa Beaumont

Patrick Burke

Dylan Furdyk

Rebekah Jamieson-Craig

Andrew Jenkins

Kate Harrison

Joe Kelly

Simon Lees

Patrick Litani

Andrew Lody

Faith Matcham

Vittorio Nigrelli

Ashley Polhemus

Sam Shephard

Sara Simblett

Kanan Tekchandani

Jack Versace

Janice Weyer

Becki Wiggett

Foreword written by Professor Barbara Wilson, OBE

I am unapologetically me.

(Ashley, Chapter 15)

This interesting book should be on every psychologist's wish list. It is, obviously, about neurodiversity as the title states, but not only does it cover neurodevelopmental conditions such as autism and ADHD, it also covers neurological conditions, including multiple sclerosis, traumatic brain injury, and mental health conditions, including schizophrenia, chronic depression, and bipolar diagnoses. Apart from Chapters 1 and 20 which explain and define neurodiversity, the core of the book is the neurodiverse people themselves as most chapters include the lived experience of 18 people. Each of these chapters begins with a brief explanation of the diagnosis, followed by an account from a neurodiverse person , and then a comment from one of the authors of the book. The first chapter gives a sensible overview of what diversity means and some of the problems faced by people with one or more of the aforementioned conditions. Some of these people face these difficulties, some embrace their differences, and some show a mixture of both. The authors discuss fluctuating conditions, too, such as MS, mental health conditions, and epilepsy. The final chapter, as well as defining accessibility (not only physical accessibility which most of us understand but also mental accessibility which is harder to understand) offers suggestions for those with neurodiversity for "No neurodivergent person has a one-size fits all way that they would want help or support. Generally, the easiest way is to ask" (D, Chapter 11). We need to educate people more about neurodiversity for it is part of the human condition, and common, but often invisible. Like all good academic books, this one is a useful source of references too. The remaining 18 chapters focus on the neurodiverse people themselves: we hear from people with ADHD, autism, epilepsy, stroke, traumatic brain injury, MS, schizophrenia, bipolar disorder, chronic depression, and tumour. All the chapters are interesting, of course, but I particularly liked Chapter 17 where we heard from Simon who had both epilepsy which was diagnosed very late and surgery. He says, "Brain differences aren't deficits: more like variations that contribute to the diversity of the human experience". Later in the chapter Simon says, "the term 'neurodivergent usually refers to an individual, whilst 'neurodiverse' is used to describe a group or a community". Chapter 8 is another chapter readers may wish to focus on: Sara explains she never used to see herself as neurodivergent but now she does. Another chapter I should point out is Chapter 9 by Becki. She was diagnosed with chronic depression, long term anxiety, and PTSD. She talks about neurodiverse people and says, "these groups are examples of people that 'deviate from the norm', but they deserve a society where their strengths, capabilities and struggles can be considered and acted upon to drive positive change for everyone". The final chapter I wish to mention is Chapter 14 about Lisa. She had a stroke, and this led to her being less cautious and less private than she was before her cerebral haemorrhage. In some ways this made her a happier person so is one example of how neurodiversity can improve life. The book certainly achieves its purpose of making the reader more aware of neurodiversity, its many manifestations, and the problems and successes people with these conditions face. I remember once hearing the great Simon Baron-Cohen speak on autism and, in the question-and-answer session, I asked about treatment. In his gentle way, Simon Baron-Cohen rebuked me saying these people do not need to be treated but embraced. He was right, of course, and since reading this book, I would not ask such a question again. The book taught me things, made me think and understand neurodiversity a little more.

Preface written by Dr Faith Matcham

In recent years, the concept of neurodiversity has gained well-deserved recognition, reshaping our understanding of human cognition and behaviour. This book is a testament to that evolution – an exploration of neurodiversity through multiple lenses, blending academic inquiry with the invaluable perspectives of those who live these experiences daily. As a health psychologist and mental health researcher, I have had the privilege of working at the intersection of psychology, technology, and lived experience, witnessing firsthand how crucial it is to honour the voices of neurodivergent individuals in research, clinical practice, and public discourse.

Neurodiversity challenges the traditional, often deficit-based models of mental and neurological health, by recognising natural variations in cognitive function as part of human diversity rather than disorders to be cured. Neurodivergent identities are not pathologies but different ways of thinking, feeling, and engaging with the world. By shifting our focus from remediation to accommodation, and from stigma to empowerment, we can create a society that embraces cognitive diversity as an asset.

As our understanding of neurodiversity evolves, so too must our approaches to support and inclusion. This requires systemic changes in education, employment, healthcare, and community spaces to ensure that neurodivergent individuals are not merely accommodated but genuinely valued. Equitable access to resources, advocacy, and policy reform must go hand in hand with shifting social attitudes to foster environments where all individuals can thrive, regardless of cognitive style.

One of the most profound shifts in this discourse has been the increasing inclusion of people with lived experience in research and policymaking. Historically, neurodivergent individuals have often been studied rather than consulted, their voices marginalised in favour of external observations. This book seeks to rectify that imbalance. Within these pages, you will find contributions from researchers, clinicians, educators, and, most importantly, neurodivergent individuals themselves. Their insights provide a richness and authenticity that no amount of external analysis could achieve alone.

This book is not just an academic resource; it is a call to action. It urges researchers to adopt participatory methods that centre neurodivergent voices. It encourages educators to design inclusive learning environments that accommodate diverse cognitive styles. It challenges clinicians to move beyond deficit-based approaches toward strength-based, affirming care. Most importantly, it invites all of us – whether neurodivergent or neurotypical – to reconsider our assumptions, to listen more deeply, and to advocate for a world where all minds are valued and supported.

As you engage with the chapters ahead, I invite you to approach them with curiosity, humility, and an openness to learning. Whether you are a researcher, clinician, educator, policymaker, or someone seeking to better understand your own neurodivergent identity, this book offers a multifaceted exploration of neurodiversity that is both intellectually rigorous and deeply human. My hope is that it will inspire meaningful dialogue, challenge outdated paradigms, and, most importantly, contribute to a more inclusive and accepting society.

Acknowledgements

Thank you to King's College London for their support in the preparation of this book.

1   Redefining neurodiversity

This opening to the book explores the definition of neurodiversity, stripping this language back to basics. In doing this the authors redefine what has been commonly thought to be the meaning of neurodiversity and neurodivergence. Throughout this chapter and the entire book, careful consideration has been given to the way we communicate about this complex and emotive topic. This chapter is an overview of the relevant scientific literature on how neurodiversity can be applied in the context of neurodevelopmental conditions, acquired neurological conditions and mental health conditions, with specific notes on functional neurological conditions and the potential fluctuations in neurodiverse experiences. Finally, the concept of 'disability' is discussed as it may or may not related to neurodiversity or neurodivergence.

Redefining neurodiversity

'Diversity' simply means difference. A diverse group of people would include a range of different experiences. Now, if we use the word 'divergent' to describe someone, we commonly mean something different from the average experience. 'Neuro' means the brain or nervous system. Putting these terms together, 'neurodiversity' means a diverse group of people with a range of differences in terms of the structure or function of the brain or nervous system. And people who are 'neurodivergent' are 'different' from average in terms of their brain or nervous system's structure or function.

Ironically, the breadth of this definition can be interpreted to include a wide range of human experiences. Perhaps the least surprising use of the term neurodivergence includes reference to people who have had a neurodevelopmental condition such as autism spectrum conditions, or attention-deficit hyperactivity disorder (ADHD) since childhood. In these cases, the brain or nervous system structure or function have developed in an atypical way.

However, a person who has been diagnosed with a neurological condition such as a brain injury, will, by definition, have experienced a change in

DOI: 10.4324/9781032714745-1

brain or nervous system structure or function. We will argue later in this chapter that all these people could identify as being neurodivergent or part of a neurodiverse group in society, and that this is an acquired difference.

It is also now widely evidenced through scientific research that people who suffer long-term mental health conditions (for example, major depressive disorder, bipolar affective disorder, or psychosis) can experience changes, maybe fluctuating changes, in brain functions such as cognition and emotion regulation. This, we will argue, means that there is reason to extend the definition of neurodiverse experiences to include those of people with cognitive differences in the context of mental health conditions.

While we first introduced the concept of being divergent as being different to average, the World Health Organization (WHO) now estimates that there are 1 billion people who have been identified as having a neurodevelopmental condition, a further 1 billion people worldwide affected by a neurological condition; and 450 million people with a long-term mental health condition. That is almost one third of the world's total population. Statistically speaking, neurodivergence is not an unusual experience and, we argue, the needs of this group deserve attention.

Side notes on language

As you read this, you may have already noticed that we have spoken about people who have 'conditions', rather than an 'illness', 'disorder', or 'disease'. This is a purposeful use of language in the attempt to adopt a term that simply indicates a state of health, whether well or ill. In this way, we remain value-neutral to be inclusive of people who live with a range of health states and functional abilities.

This leads us to explain what we understand by the terms well and ill, in the context of how we have just defined neurodiversity. Figure 1.1 shows an adapted version of the Global Wellness Institute's (2020) 'Dual Continuum Model'. One dimension represents the presence of symptoms of illness while the other dimension represents how well a person is living. It is common when following the medical model to see distress in terms of the presence of symptoms that require treatment to reduce an underlying pathology. However, in the Dual Continuum Model it is acknowledged that people can and do often live well even if symptoms of illness are present. High wellness, regardless of the illness state, is conceptualised as 'flourishing', and originates from the field of Positive Psychology.

Holding this model in mind, someone who fits our definition of being neurodivergent may have differences in the way their brain or nervous symptom is structured and is functioning but can be capable of being very well, living a happy, satisfied life.

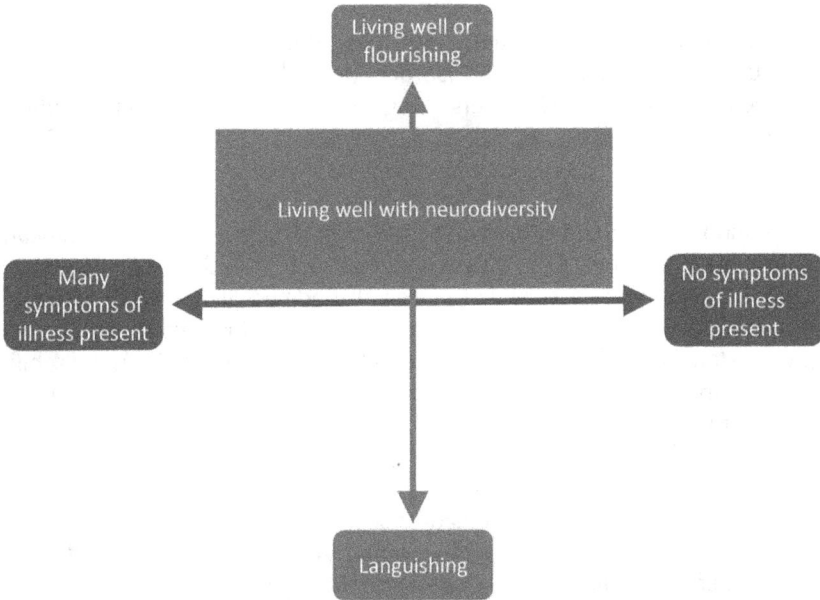

Figure 1.1 An adaptation to the Global Wellness Institute's (2020) 'Dual Continuum Model' to represent living well with neurodiversity

Applications of neurodiversity

The next section of this book will be about understanding how our definition of neurodiversity and neurodivergence applies to neurodevelopmental conditions, then neurological conditions, and, lastly, mental health conditions.

Neurodiversity applied to neurodevelopmental conditions

Before describing how the definition of neurodiversity applies to neurode-velopmental conditions, we must first define what is meant by that term. In two of the most frequently used diagnostic frameworks (DSM-5, the Diagnostic and Statistical Manual and ICD-11, the International Classification of Diseases), neurodevelopmental conditions are broadly defined as a group of conditions with onset in the developmental period (i.e., in childhood, specifically before the age of 12), causing 'clinically significant' (DSM-5) or 'sufficiently severe' (ICD-11) impairments of functioning (e.g., academically, occupationally, or socially). Neurodevelopmental conditions can include (according to Bishop & Rutter, 2008):

1 Congenital intellectual disabilities (ID), including neurogenetic conditions such as Fragile X, Down's Syndrome
2 Communication or Speech and Language Disorders

Note the use of the term 'disorder' in all cases, and the lack of guidance in the DSM-5 and ICD-11 about quantifying severity or defining 'clinically significant' impairments in function. Also, please note that such 'disorders' may not exist in isolation. For example, according to research findings, 50% to 70% of people with ASD also meet diagnostic criteria for ADHD (Hours et al., 2022), and approximately 20% of people with ASD will also have an intellectual disability (Khachadourian et al., 2023). Genetic research suggests that ID, ASD, and ADHD (as well as some mental health conditions, i.e., bipolar affective disorder and psychotic disorders, which we will come to later in this chapter) all lie on the same neurodevelopmental continuum (Morris-Rosendahl & Crocq, 2020). Hours et al. argue that instead of seeing the presence of multiple conditions as 'comorbidities' we should be trying to understand underlying 'traits', such as differences with cognitive (or brain-related) functions.

What does this mean in terms of neurodiversity? It means that there are certain biological mechanisms that underlie these conditions and make them like each other in some way. The premise of this book is that these similarities can be traced back to brain or nervous system structure or function. Let's take people with a diagnosis of ADHD; meta-analyses of neuroimaging studies have shown structural differences in the brain of people with this diagnosis (Frodl & Skokauskas, 2012; Lukito et al., 2020). Similarly, people with ASD have been found to have structural brain differences, albeit affecting different regions of the brain to people with ADHD (Lukito et al., 2020). It is also true that structure brain changes have been found for people with ID (Spencer et al., 2006), communication and speech and language disorders (Badcock et al., 2012), neurodevelopmental motor disorders (Perrotta, 2019), and specific learning disorders (Yan et al., 2021).

You may be asking yourself the question, what are the consequences of these structural brain differences? One answer is that people with neurodevelopmental conditions may have observable differences in their emotional, cognitive, and sensory processing, which in turn may affect their behaviour. Table 1.1 provides some examples from research findings.

This is not to say that people with neurodevelopmental conditions cannot live well. On the contrary, happiness (Stickley et al., 2018), life satisfaction (Franke et al., 2019), strong relationships (Marton et al., 2015; Tipton et al., 2013), personal growth, and 'peak' performance (in some cases 'giftedness'; Kontakou et al., 2022) can be just as present for people with who are either diagnosed with and or identify as living with a neurodevelopmental condition. This means that an individual can, with the right environment and support, still flourish in terms of their wellbeing.

Table 1.1 Examples of evidence-based emotional, cognitive, sensory, and behavioural differences for people diagnosed with a neurodevelopmental condition

Neurodevelopmental condition	Emotional, cognitive, sensory, and behavioural differences	References
Intellectual Disabilities (ID)	↓ Planning ↓ Problem-solving ↓ Decision making ↓ Memory ↓ Reasoning	Lifshitz et al. (2016) Palomino Plaza et al. (2019) Spaniol & Danielsson (2022)
Communication or Speech and Language Disorders	↓ Speech production ↓ Speech perception ↓ Reasoning (including non-verbal) ↓ Motor skills	Estes et al. (2007) Hearnshaw et al. (2019) Gallinat and Spaulding (2014) Rechetnikov and Maitra (2009)
Autism Spectrum Disorders (ASD)	↑ Sensitivity to auditory information ↓ Recognising emotions ↓ Social cognition	Carpenter and Williams (2023) Williams et al. (2021) Velikonja et al. (2019)
Attention-Deficit Hyperactivity Disorder (ADHD)	↓ Decision making ↓ Impulse/executive control ↓ Concentration ↓ Verbal working memory ↓ Planning	Dekkers et al. (2021) Patros et al. (2019) Schoechlin and Engel (2005) Ramos et al. (2020)
Neurodevelopmental Motor Disorders (e.g., tic disorder or Tourette Syndrome)	↓ Impulse/executive control ↓ Social cognition	Cavanna et al. (2020)
Specific Learning Disorders (e.g., dyslexia)	↓ Sustained attention ↓ Working memory	de Assis Leão et al. (2023) Reis et al. (2020)

But how about those individuals without the right environment and support who may not be flourishing? Some people with neurodevelopmental conditions have been found to report lower quality of life (Barneveld et al., 2014). Moderators include poor mental health such as depression, anxiety, or low self-esteem (Smith et al., 2019; Tsermentseli, 2022), and challenges within a person's family and wider support system including financial pressures, unequal access to healthcare, and high levels of dependency on others (Bishop-Fitzpatrick et al., 2018; Ueda et al., 2022). The list could go on, but we will leave it there. The key message, and probably a relatively uncontroversial one, is that people with neurodevelopmental conditions are considered neurodivergent under our definition and this will come with its strengths but, at times, perhaps where support is lacking, difficulties. In later chapters, we will hear first-hand experiences that will provide an insight into the reality of what it is like to live with a neurodevelopmental condition in our current society. For now, let us focus on the breadth of our new definition of neurodiversity.

Neurodiversity applied to acquired neurological conditions

The term neurological conditions cover a vast range of diagnoses, and it is perhaps the easiest in which to evidence differences in the brain or nervous system's structure or function, as this is a required feature. Some of the key conditions that we will refer to in this chapter and later in the book include:

1   Non-progressive acquired brain injury – ABI (including traumatic brain injury; brain-related inflammatory and autoimmune conditions such as encephalitis; hypoxia, e.g., due to cardiovascular events; brain tumours; and stroke or cerebrovascular accidents)
2   Epilepsy
3   Other autoimmune conditions (including MS (multiple sclerosis) and lupus)
4   Neurodegenerative or progressive conditions (including PD (Parkinson's disease), HD (Huntingdon's disease), and all dementias)
5   Acquired neurodevelopmental conditions (including ABI in childhood, PKU (Phenylketonuria), TSC (tuberous sclerosis complex))

The nature of differences in the brain or nervous system's structure or function for people with ABI (regardless of age) depends on the type of acquired injury or illness. Take traumatic brain injury (TBI), for example, many people suffer this type of injury because of a road traffic accident (RTA). The Mayo Classification of TBI severity states that people with moderate or severe (definite) TBI will have one of the following present: an intracerebral, subdural, or epidural hematoma, a cerebral or haemorrhagic contusion, a subarachnoid haemorrhage, or a penetrating or brain stem injury. Further evidence of post-traumatic amnesia (PTA), loss of consciousness (LOC), and the Glasgow Coma Scale (GCS) score during the first 24 hours make up the full criteria, but they are beyond the scope of this book, which focuses on neurodiversity in the context of brain or nervous system's structure or function. People with TBI commonly present with cognitive difficulties with executive functioning (i.e., planning, problem-solving, decision-making, emotional and behaviour self-regulation, initiation and drive, social cognition, and self-awareness) that relate to changes in brain structure or function of the frontal lobe (Stuss, 2011). Further difficulties with processing speed may be a result of diffuse axonal injury (Felmingham et al., 2004). ABIs caused by strokes can be more varied in their effects on the brain. Subtypes are dependent on the region of the blood vessels affected and where they usually supply oxygen to the brain, and extent of the ischemia (blockage of a blood vessel) or haemorrhage (bleed in the brain) (Amarenco et al., 2009). This is similarly the case for people with brain tumours or tubers in TSC.

Different neurodegenerative or progressive conditions affect specific regions of the brain. For example, in PD there is pathology found in an area of the brain called the 'basal ganglia' that has a systemic effect on nervous system function (Schneider, 2014). Different types of dementia result in brain atrophy in different regions of the brain and consequences of functioning (Gure et al., 2010). Some neurological conditions fluctuate in terms of their effects on the brain and functional presentation, for example, epilepsy and MS (relapsing-remitting subtype), and thereby cause a temporary difference in brain or nervous system's structure or function. In the case of MS, relapses lead to changes in the myelin sheath surrounding the neurons in the brain that can sometimes repair over time (Podbielska et al., 2013). In the case of epilepsy, seizures caused by a sudden, uncontrolled burst of electrical activity in the brain interrupt brain functioning and can be reversible (Aldenkamp et al., 2005).

The type of difficulties that result is somewhat irrelevant to the context of this book of which the premise is that anyone with differences in the brain or nervous system's structure or function, be it permanent or temporary, is, according to the reframed definition, neurodivergent.

A note on functional neurological conditions

Perhaps one 'category' of conditions that should be included in our new definition of neurodivergence are 'functional neurological conditions' or 'disorders' (commonly referred to as FND). The very definition is a behaviour or sensation that is altered without a structural change in the brain or central nervous system, hence a 'functional' problem. However, to conclude that the brain is not 'involved' in these conditions is perhaps misleading. Functional magnetic resonance imagining (fMRI) studies have revealed differences in the way blood flows around the brain in people with FND, more specifically increased corticolimbic activity and reduced activity in the right temporoparietal junction (Perez et al., 2021). For some, these changes in brain function may manifest in cognitive problems. While subjective problems with cognition are very common, empirical evidence for objective impairment in neurocognitive functioning is variable but, where supported, indicates differences in attention, executive functioning, and social cognition (Pick et al., 2023).

Neurodiversity applied to mental health conditions

The literature on mental health conditions is vast, and while neurodivergence as applied to neurodevelopmental conditions may be seen as relatively uncontroversial, the same cannot perhaps be said for this section of the book. Because what we mean by mental health conditions is vast, we will try to focus our definition to encompass the experiences where there is the greatest evidence of, as our definition of neurodivergence states, differences in terms of the structure or function of the brain or nervous system:

1   Major Depressive Disorder (MDD)
2   Bipolar Affective Disorder (BPAD)
3   Psychotic conditions (e.g., 'schizophrenia')
4   Stress-related and anxiety disorders (including PTSD (post-traumatic stress disorder), social phobia and other specific phobias, GAD (generalised anxiety disorder), OCD (obsessive-compulsive disorder), panic disorder with and without agoraphobia)

As with neurodevelopmental conditions, there can be some overlap in the experience of these mental health conditions. Also, as previously referenced, they (specifically BPAD and psychosis) may lie on a genetic continuum with ID, ASD, and ADHD (Morris-Rosendahl & Crocq, 2020). Let us, however, begin with a discussion of MDD. Meta-analyses (some from over a decade ago) have revealed consistent structural abnormalities in the fronto-limbic neural networks (i.e., connections between cortical and subcortical regions, including the basal ganglia and hippocampus; these parts of the brain play a crucial role in the processing of 'reward' and 'episodic memory', respectively) are the essential characteristics of MDD in adulthood (Lorenzetti et al., 2009; Peng et al., 2016).These findings have been replicated in adolescents with depression (Shen et al., 2021).

Alongside these structural differences, neuroimaging studies focusing on brain 'function' found significantly reduced striatal activation (part of the basal ganglia) in depressed individuals during reward tasks compared to controls. This effect was seen particularly during reward anticipation, with a more substantial effect in individuals under 18 (Keren et al., 2018). What makes this topic area complex is that these structural and functional brain changes may be temporary in nature. Combined findings from multiple studies have found that pharmacotherapy and psychological treatment can result in improvements in brain structure (e.g., connectivity) and function (Delaveau et al., 2011; Nord et al., 2021).

We must remember that these are group-based findings, and they do not prove that everyone with MDD has either structural or functional brain changes, but they do indicate that the concept of neurodivergence, as presented in this book, could be relevant. Taking a more neuropsychological approach, one study has revealed that over 50% of people with MDD demonstrated a significant level of cognitive impairment, with this reducing to 24% after antidepressant treatment, with problems in attention and executive functioning more commonly remaining in this group (Guo et al., 2023). Another longitudinal study showed that the more persistent the cognitive difficulties the higher the levels of depression and functional impairment (Matcham et al., 2023).

Although BPAD and psychotic conditions make up a smaller group of people with mental health conditions, they are particularly relevant to the subject of this book due to a deep literature that has emerged focusing on the cognitive, and thus brain-related, needs of these people. It is now widely accepted that people with a diagnosis of psychotic conditions such as 'schizophrenia' have clearly observable cognitive deficits (McCutcheon et al., 2023). The trajectory of these cognitive difficulties does not appear to change significantly over time (Watson et al., 2022) and it is not a novel idea that schizophrenia could be seen as a neurodevelopmental condition due to cognitive difficulties preceding the onset of psychosis (Murray & Lewis, 1987). Cognitive Remediation Therapy (CRT) is now the focus of treatment for longstanding brain-related effects of psychotic conditions (Cella et al., 2020; Wykes et al., 2011), due to mixed and inconclusive evidence that pharmacotherapy (i.e., antipsychotic medication) can make an improvement to cognitive functioning (Anda et al., 2021).

When it comes to people presenting with mania in BPAD, group-based structural changes have been found in the prefrontal cortex of the brain (Abé et al., 2023), which may explain problems with executive functioning, and specifically impulse control (Swann et al., 2009). There is greater evidence of improvement of cognitive function in between episodes of illness in BPAD, but still, approximately one-third of people report ongoing problems when 'euthymic' (i.e., when not experiencing mood-related symptoms) (Tsapekos et al., 2021).

Attentional bias, for example, towards the source or interpretation of a threat, is a common feature of anxiety and stress-related disorders. So much so that there is now mounting evidence that interventions that target these mechanisms and seek to alter the focus of attention or interpretation away from the threat, i.e., Cognitive Bias Modification (CBM), show a small but consistent effect that could play a role in clinical treatment of these disorders (Fodor et al., 2020).

While it is perhaps not common to label people with anxiety disorder as having 'cognitive difficulties', the effectiveness of interventions such as CBM suggests otherwise. When you delve a little deeper into the research findings, cognitive difficulties have been found to be present. For example, difficulties with episodic memory and executive functioning, are mostly experienced by people with panic disorder with and without agoraphobia, and obsessive–compulsive disorder (Airaksinen et al., 2005). There is growing evidence of the role of cognitive difficulties in PTSD, with the largest effect sizes in verbal learning, speed of information processing, attention and or working memory, and verbal memory (Scott et al., 2015). Problems with social cognition (also referred to as 'mentalising') can be experienced by people with social phobias (Ballespí et al., 2021). Note the overlap in difficulties with people who have a diagnosis of neurodevelopmental conditions, such as ASD. It may also be important to note that difficulties with mentalising may be a transdiagnostic cognitive process across other mental health diagnoses, including psychotic disorders such as schizophrenia, BPAD, and PTSD (Bora et al., 2016; Stevens & Jovanovic, 2019). Another process thought to be common across anxiety disorders, MDD, and other mental health conditions is repetitive negative thinking (e.g., rumination and worry), which has been found to be associated with cognitive control processes or working memory (Zetsche et al., 2018).

Taking these findings together, we hope the reader of this book will be able to understand why we feel strongly that mental health conditions should be considered in the context of neurodiversity. Again, as with the people with the neurodevelopmental conditions previously discussed, we are aware that many people are living well with perhaps a 'history' or at least infrequent episodes of mental illness. People reporting high levels of quality of life despite a history of mental illness have endorsed the feeling of being in control (particularly of distressing symptoms), having a sense of autonomy, choice, a positive self-image, belonging, and were engaging in meaningful and enjoyable activities, feeling hope, and optimism (Connell et al., 2012).

Fluctuations in the neurodiversity spectrum

The possible temporary nature of brain and nervous system structural and or functional changes has already been acknowledged. Some neurological conditions, for example epilepsy and autoimmune disorders such as relapsing remitting MS and lupus, and most mental health conditions, are episodic in nature. This means that people with these conditions may experience a difference in the functioning or structure of the brain or nervous system at some times but not others, and the severity of these differences may vary. Never has it been more important to conceptualise neurodiversity as a spectrum as when we think about such fluctuations. It is interesting to note whether people who experience fluctuating health conditions would even identify with the term 'neurodiversity', as it has perhaps, more traditionally, been conceptualised in a static way. However, just because a person may not have the same needs in every moment of their life, does not mean that we, as a society, should not be thinking of supporting these individuals with sometimes necessary accommodations. This will be a prominent theme explored in the lived experience chapters later in this book.

Another key message

Something to take away from this section of the book is the high level of overlap between the presentation of people discussed under the subtitles of 'neurodevelopmental conditions', 'mental health conditions', and 'neurological conditions'. All have been shown to be associated with cognitive differences, and sometimes it is hard to decide which condition label should be given, as is clearly the case for people with acquired neurodevelopmental conditions where this leads to intellectual disabilities. The authors believe that this serves to strengthen the argument and premise of this book, that neurodiversity is a wide range and broad spectrum of sometimes overlapping conditions.

Disability and how it relates to neurodiversity

We will now turn to another thorny and somewhat controversial issue and that is whether people who are neurodivergent do or do not have a 'neuro-disability.' A disability under the Equality Act 2010 refers to a physical or mental impairment that has a 'substantial' and 'long-term' negative effect on a person's ability to do normal daily activities. We have already established that most people who we have already defined as neurodivergent will have a difference in terms of the structure or function of the brain or nervous system that results in a change in cognitive, emotional, or behavioural functioning from people considered to be 'neurotypical'. However, the extent to which this change can be labelled an 'impairment' is what makes this a complex subject.

Take someone who has had a TBI, who prior to the road traffic accident that they were involved in had a general IQ as well as more specific cognitive functions in the high average range. A change in cognitive functioning could mean that this individual's cognitive function in some areas drops to the average range. This means that this person is now functioning in line with the majority (50%) of the population. But this is only because the person had a high level of 'cognitive reserve'; they may experience the change as a difference and in many cases limiting to the life they had prior to the accident. Even if this person identifies as having a disability now, does this count under the law as a mental impairment?

Even the second part of the definition of disability is ambiguous: 'substantial' is more than a minor or trivial impairment. They provide an example of it taking much longer than it usually would to complete a daily task like getting dressed. But, again, what if the person prior to an injury or illness was functioning as an extremely high level, e.g., an executive position in a company – their concept of what daily tasks they need to complete would potentially be greater than others, to the extent that if they can no longer carry out their executive-level duties this drop in functioning could be experienced as 'substantial' even if they are able to complete a less demanding role. What is perhaps missing from our current definition of disability is the relative nature of impairment for people with acquired limitations.

Then we must consider the meaning of a 'long-term' condition. How long is 'long-term' and what about conditions where there is a fluctuating course, such as in the case of relapsing remitting multiple sclerosis, most mental health conditions, and epilepsy? Perhaps we need to take note of the fact that many of these people take medication to stop symptoms from manifesting, and may, in some cases, prevent the expression of 'illnesses' entirely. If following The Global Wellness Institute's (2020) 'Dual Continuum Model', one might argue that, regardless of the presence of symptoms of 'illness', it is those who are languishing 'long-term' who have a 'disability', if 'disability' is to mean 'disadvantage' or 'limitation'. But then, what about all the people who are flourishing in spite of everything? The person with autism who has found their niche working in an area of specialist interest, or the person who makes it their goal in life to be a 'lived experience expert' in their condition and help shape services and products to better meet the needs of people with neurocognitive conditions? Does the concept of 'disability' describe these people's experiences? In fact, there may be a multitude of reasons why a person chooses not to identify as a person with a disability, including fear of stigma and discrimination, or a personal choice to focus on strengths rather than limitations.

Adopting this reframed model of neurodiversity

In adopting our reframed model of neurodiversity, we are not claiming that all people with a diagnosis of a neurological, mental health or neurodevelopmental condition necessarily have a disability nor necessarily have limitations in terms of their ability to function in everyday life. Neither are we saying that we should ignore or inflate neurodivergence above other dimensions of diversity, and we very much acknowledge the complex intersectionality between the experiences of people who identify as neurodivergent and 'minoritised' for their age, sexual orientation, gender, class, race, religious beliefs, and / or other protected characteristics. We are merely revisiting a definition that we hope will go some way to meeting the needs of all people with needs that result from a difference in the brain or nervous system's structure or function. Finally, though, we acknowledge fully that these needs will likely not be static. Supporting people with neurodiverse needs will be a continual conversation and evolving process and requires each person to be treated as an individual.

In the next section of this book, we will take a step back from the theory and a step into the lives of people living with neurodiversity. In each chapter we will hear from a single person why they do (or do not) identify with the new definition we propose. We will learn what needs each of these people have and how these needs are and are not currently met in society. Finally, we will learn from each individual what they hope will change if neurodiversity was better understood.

References

Abé, C., Liberg, B., Klahn, A. L., Petrovic, P., & Landén, M. (2023). Mania-related effects on structural brain changes in bipolar disorder–a narrative review of the evidence. *Molecular Psychiatry*, 28(7), 2674–2682.

Airaksinen, E., Larsson, M., & Forsell, Y. (2005). Neuropsychological functions in anxiety disorders in population-based samples: evidence of episodic memory dysfunction. *Journal of Psychiactric Research*, 39(2), 207–214.

Aldenkamp, A. P., Beitler, J., Arends, J., van der Linden, I., & Diepman, L. (2005). Acute effects of subclinical epileptiform EEG discharges on cognitive activation. *Functional neurology*, 20(1), 23–28.

Amarenco, P., Bogousslavsky, J., Caplan, L., Donnan, G., & Hennerici, M. (2009). Classification of stroke subtypes. *Cerebrovascular diseases*, 27(5), 493–501.

Anda, L., Johnsen, E., Kroken, R. A., Joa, I., Rettenbacher, M., & Løberg, E.-M. (2021). Cognitive change and antipsychotic medications: Results from a pragmatic rater-blind RCT. *Schizophrenia Research: Cognition*, 26, 100204.

Badcock, N. A., Bishop, D. V., Hardiman, M. J., Barry, J. G., & Watkins, K. E. (2012). Co-localisation of abnormal brain structure and function in specific language impairment. *Brain and Language*, 120(3), 310–320.

Ballespí, S., Vives, J., Nonweiler, J., Perez-Domingo, A., & Barrantes-Vidal, N. (2021). Self-but not other-dimensions of mentalizing moderate the impairment associated with social anxiety in adolescents from the general population. *Frontiers in Psychology*, 12, 721584.

Barneveld, P. S., Swaab, H., Fagel, S., Van Engeland, H., & de Sonneville, L. M. (2014). Quality of life: A case-controlled long-term follow-up study, comparing young high-functioning adults with autism spectrum disorders with adults with other psychiatric disorders diagnosed in childhood. *Comprehensive Psychiatry*, 55(2), 302–310.

Bishop, D., & Rutter, M. (2008). Neurodevelopmental disorders: conceptual issues. *Rutter's Child and Adolescent Psychiatry*, 5, 32–41.

Bishop-Fitzpatrick, L., Mazefsky, C. A., & Eack, S. M. (2018). The combined impact of social support and perceived stress on quality of life in adults with autism spectrum disorder and without intellectual disability. *Autism*, 22(6), 703–711.

Bora, E., Bartholomeusz, C., & Pantelis, C. (2016). Meta-analysis of Theory of Mind (ToM) impairment in bipolar disorder. *Psychological medicine*, 46(2), 253–264.

Carpenter, K. L., & Williams, D. M. (2023). A meta-analysis and critical review of metacognitive accuracy in autism. *Autism*, 27(2), 512–525.

Cavanna, A. E., Ganos, C., Hartmann, A., Martino, D., Pringsheim, T., & Seri, S. (2020). The cognitive neuropsychiatry of Tourette syndrome. *Cognitive Neuropsychiatry*, 25(4), 254–268.

Cella, M., Price, T., Corboy, H., Onwumere, J., Shergill, S., & Preti, A. (2020).-Cognitive remediation for inpatients with psychosis: a systematic review and meta-analysis. *Psychological Medicine*, 50(7), 1062–1076.

Connell, J., Brazier, J., O'Cathain, A., Lloyd-Jones, M., & Paisley, S. (2012). Quality of life of people with mental health problems: a synthesis of qualitative research. *Health and Quality of Life Outcomes*, 10, 1–16.

de Assis Leão, S. E. S., Menezes Lage, G., Pedra de Souza, R., Holanda Marinho Nogueira, N. G. d., & Vieira Pinheiro, Â. M. (2023). Working Memory and Manual Dexterity in Dyslexic Children: A Systematic Review and Meta-Analysis. *Developmental Neuropsychology*, 48(1), 1–30.

Dekkers, T. J., Agelink van Rentergem, J. A., Huizenga, H. M., Raber, H., Shoham, R., Popma, A., & Pollak, Y. (2021). Decision-making deficits in ADHD are not related to risk seeking but to suboptimal decision-making: Meta-analytical and novel experimental evidence. *Journal of Attention Disorders*, 25(4), 486–501.

Delaveau, P., Jabourian, M., Lemogne, C., Guionnet, S., Bergouignan, L., & Fossati, P. (2011). Brain effects of antidepressants in major depression: a meta-analysis of emotional processing studies. *Journal of Affective Disorders*, 130(1–2), 66–74.

Estes, K. G., Evans, J. L., & Else-Quest, N. M. (2007). Differences in the nonword repetition performance of children with and without specific language impairment: A meta-analysis. *Journal of Speech, Language, and Hearing Research*, 50 (1), 177–195.

Felmingham, K. L., Baguley, I. J., & Green, A. M. (2004). Effects of diffuse axonal injury on speed of information processing following severe traumatic brain injury. *Neuropsychology*, 18(3), 564.

Fodor, L. A., Georgescu, R., Cuijpers, P., Szamoskozi, Ş., David, D., Furukawa, T. A., & Cristea, I. A. (2020). Efficacy of cognitive bias modification interventions in anxiety and depressive disorders: a systematic review and network meta-analysis. *The Lancet Psychiatry*, 7(6), 506–514.

Franke, K. B., Hills, K., Huebner, E. S., & Flory, K. (2019). Life satisfaction in adolescents with autism spectrum disorder. *Journal of Autism and Developmental Disorders*, 49, 1205–1218.

Frodl, T., & Skokauskas, N. (2012). Meta-analysis of structural MRI studies in children and adults with attention deficit hyperactivity disorder indicates treatment effects. *Acta Psychiatrica Scandinavica*, 125(2), 114–126.

Gallinat, E., & Spaulding, T. J. (2014). Differences in the performance of children with specific language impairment and their typically developing peers on nonverbal cognitive tests: A meta-analysis. *Journal of Speech, Language, and Hearing Research*, 57(4), 1363–1382.

Guo, W., Liu, B., Wei, X., Ju, Y., Wang, M., Dong, Q., Lu, X., Sun, J., Zhang, L., & Guo, H. (2023). The longitudinal change pattern of cognitive subtypes in medication-free patients with major depressive disorder: a cluster analysis. *Psychiatry Research*, 327, 115413.

Gure, T. R., Kabeto, M. U., Plassman, B. L., Piette, J. D., & Langa, K. M. (2010). Differences in functional impairment across subtypes of dementia. *Journals of Gerontology Series A: Biomedical Sciences and Medical Sciences*, 65(4), 434–441.

Hearnshaw, S., Baker, E., & Munro, N. (2019). Speech perception skills of children with speech sound disorders: A systematic review and meta-analysis. *Journal of Speech, Language, and Hearing Research*, 62(10), 3771–3789.

Hours, C., Recasens, C., & Baleyte, J.-M. (2022). ASD and ADHD comorbidity: What are we talking about? *Frontiers in Psychiatry*, 13, 837424.

Keren, H., O'Callaghan, G., Vidal-Ribas, P., Buzzell, G. A., Brotman, M. A., Leibenluft, E., Pan, P. M., Meffert, L., Kaiser, A., & Wolke, S. (2018). Reward processing in depression: a conceptual and meta-analytic review across fMRI and EEG studies. *American Journal of Psychiatry*, 175(11), 1111–1120.

Khachadourian, V., Mahjani, B., Sandin, S., Kolevzon, A., Buxbaum, J. D., Reichenberg, A., & Janecka, M. (2023). Comorbidities in autism spectrum disorder and their etiologies. *Translational Psychiatry*, 13(1), 71.

Kontakou, A., Dimitriou, G., Panagouli, E., Thomaidis, L., Psaltopoulou, T., Sergentanis, T. N., & Tsitsika, A. (2022). Giftedness and neurodevelopmental disorders in children and adolescents: A systematic review. *Journal of Developmental & Behavioral Pediatrics*, 43(7), e483–e497.

Lifshitz, H., Kilberg, E., & Vakil, E. (2016). Working memory studies among individuals with intellectual disability: An integrative research review. *Research in Developmental Disabilities*, 59, 147–165.

Lorenzetti, V., Allen, N. B., Fornito, A., & Yücel, M. (2009). Structural brain abnormalities in major depressive disorder: a selective review of recent MRI studies. *Journal of affective Disorders*, 117(1–2), 1–17.

Lukito, S., Norman, L., Carlisi, C., Radua, J., Hart, H., Simonoff, E., & Rubia, K. (2020). Comparative meta-analyses of brain structural and functional abnormalities during cognitive control in attention-deficit/hyperactivity disorder and autism spectrum disorder. *Psychological medicine*, 50(6), 894–919.

Marton, I., Wiener, J., Rogers, M., & Moore, C. (2015). Friendship characteristics of children with ADHD. *Journal of Attention Disorders*, 19(10), 872–881.

Matcham, F., Simblett, S., Leightley, D., Dalby, M., Siddi, S., Haro, J. M., Lamers, F., Penninx, B. W., Bruce, S., & Nica, R. (2023). The association between persistent cognitive difficulties and depression and functional outcomes in people with major depressive disorder. *Psychological Medicine*, 53(13), 6334–6344.

McCutcheon, R. A., Keefe, R. S., & McGuire, P. K. (2023). Cognitive impairment in schizophrenia: aetiology, pathophysiology, and treatment. *Molecular Psychiatry*, 28(5), 1902–1918.

Morris-Rosendahl, D. J. & Crocq, M.-A. (2020). Neurodevelopmental disorders—the history and future of a diagnostic concept. *Dialogues in Clinical Neuroscience*, 22(1), 65–72.

Murray, R. M. & Lewis, S. W. (1987). Is schizophrenia a neurodevelopmental disorder? *British Medical Journal (Clinical Research Ed.)*, 295(6600), 681.

Nord, C. L., Barrett, L. F., Lindquist, K. A., Ma, Y., Marwood, L., Satpute, A. B., & Dalgleish, T. (2021). Neural effects of antidepressant medication and psychological treatments: a quantitative synthesis across three meta-analyses. *The British Journal of Psychiatry*, 219(4), 546–550.

Palomino Plaza, E., López Frutos, J. M., Botella Ausina, J., & Sotillo Méndez, M. (2019). Impairment of cognitive memory inhibition in individuals with intellectual disability: a meta-analysis. *Psicothema*, 31(4), 384.

Patros, C. H., Tarle, S. J., Alderson, R. M., Lea, S. E., & Arrington, E. F. (2019). Planning deficits in children with attention-deficit/hyperactivity disorder (ADHD): A meta-analytic review of tower task performance. *Neuropsychology*, 33(3), 425.

Peng, W., Chen, Z., Yin, L., Jia, Z., & Gong, Q. (2016). Essential brain structural alterations in major depressive disorder: a voxel-wise meta-analysis on first episode, medication-naive patients. *Journal of affective disorders*, 199, 114–123.

Perez, D. L., Nicholson, T. R., Asadi-Pooya, A. A., Bègue, I., Butler, M., Carson, A. J., David, A. S., Deeley, Q., Diez, I., & Edwards, M. J. (2021). Neuroimaging in functional neurological disorder: state of the field and research agenda. *NeuroImage: Clinical*, 30, 102623.

Perrotta, G. (2019). Tic disorder: definition, clinical contexts, differential diagnosis, neural correlates and therapeutic approaches. *Journal of Rehabilitation Neurosciences*, 2019, 1–6.

Pick, S., Millman, L. M., Sun, Y., Short, E., Stanton, B., Winston, J. S., Mehta, M. A., Nicholson, T. R., Reinders, A. A., & David, A. S. (2023). Objective and subjective neurocognitive functioning in functional motor symptoms and functional seizures: preliminary findings. *Journal of Clinical and Experimental Neuropsychology*, 45 (10), 970–987.

Podbielska, M., Banik, N. L., Kurowska, E., & Hogan, E. L. (2013). Myelin recovery in multiple sclerosis: the challenge of remyelination. *Brain Sciences*, 3(3), 1282–1324.

Ramos, A. A., Hamdan, A. C., & Machado, L. (2020). A meta-analysis on verbal working memory in children and adolescents with ADHD. *The Clinical Neuropsychologist*, 34(5), 873–898.

Rechetnikov, R. P. & Maitra, K. (2009). Motor impairments in children associated with impairments of speech or language: A meta-analytic review of research literature. *The American Journal of Occupational Therapy*, 63(3), 255–263.

Reis, A., Araújo, S., Morais, I. S., & Faísca, L. (2020). Reading and reading-related skills in adults with dyslexia from different orthographic systems: A review and meta-analysis. *Annals of Dyslexia*, 70(3), 339–368.

Schneider, J. (2014). Gangliosides and glycolipids in neurodegenerative disorders. *Glycobiology of the Nervous System*, 449–461.

Schoechlin, C., & Engel, R. R. (2005). Neuropsychological performance in adult attention-deficit hyperactivity disorder: Meta-analysis of empirical data. *Archives of Clinical Neuropsychology*, 20(6), 727–744.

Scott, J. C., Matt, G. E., Wrocklage, K. M., Crnich, C., Jordan, J., Southwick, S. M., Krystal, J. H., & Schweinsburg, B. C. (2015). A quantitative meta-analysis of neurocognitive functioning in posttraumatic stress disorder. *Psychological Bulletin*, 141(1), 105.

Shen, X., MacSweeney, N., Chan, S. W., Barbu, M. C., Adams, M. J., Lawrie, S. M., Romaniuk, L., McIntosh, A. M., & Whalley, H. C. (2021). Brain structural associations with depression in a large early adolescent sample (the ABCD study®). *EClinicalMedicine*, 42.

Smith, I. C., Ollendick, T. H., & White, S. W. (2019). Anxiety moderates the influence of ASD severity on quality of life in adults with ASD. *Research in Autism Spectrum Disorders*, 62, 39–47.

Spaniol, M. & Danielsson, H. (2022). A meta-analysis of the executive function components inhibition, shifting, and attention in intellectual disabilities. *Journal of Intellectual Disability Research*, 66(1–2),9–31.

Spencer, M. D., Moorhead, T. W. J., Lymer, G. K. S., Job, D. E., Muir, W. J., Hoare, P., Owens, D. G., Lawrie, S. M., & Johnstone, E. C. (2006). Structural correlates of intellectual impairment and autistic features in adolescents. *Neuroimage*, 33(4), 1136–1144.

Stevens, J. S. & Jovanovic, T. (2019). Role of social cognition in post-traumatic stress disorder: A review and meta-analysis. *Genes, Brain and Behavior*, 18(1), e12518.

Stickley, A., Koyanagi, A., Takahashi, H., Ruchkin, V., Inoue, Y., Yazawa, A., & Kamio, Y. (2018). Attention-deficit/hyperactivity disorder symptoms and happiness among adults in the general population. *Psychiatry Research*, 265, 317–323.

Stuss, D. T. (2011). Functions of the frontal lobes: relation to executive functions. *Journal of the international neuropsychological Society*, 17(5), 759–765.

Swann, A. C., Lijffijt, M., Lane, S. D., Steinberg, J. L., & Moeller, F. G. (2009). Increased trait-like impulsivity and course of illness in bipolar disorder. *Bipolar Disorders*, 11(3), 280–288.

Tipton, L. A., Christensen, L., & Blacher, J. (2013). Friendship quality in adolescents with and without an intellectual disability. *Journal of Applied Research in Intellectual Disabilities*, 26(6), 522–532.

Tsapekos, D., Strawbridge, R., Cella, M., Wykes, T., & Young, A. H. (2021). Cognitive impairment in euthymic patients with bipolar disorder: prevalence estimation and model selection for predictors of cognitive performance. *Journal of Affective Disorders*, 294, 497–504.

Tsermentseli, S. (2022). Self-esteem moderates the impact of perceived social support on the life satisfaction of adults with autism spectrum disorder. *Autism & Developmental Language Impairments*, 7. doi:23969415221147430.

Ueda, R., Okada, T., Kita, Y., Ukezono, M., Takada, M., Ozawa, Y., Inoue, H., Shioda, M., Kono, Y., & Kono, C. (2022). Quality of life of children with neurodevelopmental disorders and their parents during the COVID-19 pandemic: a 1-year follow-up study. *Scientific Reports*, 12(1), 4298.

Velikonja, T., Fett, A.-K., & Velthorst, E. (2019). Patterns of nonsocial and social cognitive functioning in adults with autism spectrum disorder: A systematic review and meta-analysis. *JAMA Psychiatry*, 76(2), 135–151.

Watson, A. J., Harrison, L., Preti, A., Wykes, T., & Cella, M. (2022). Cognitive trajectories following onset of psychosis: a meta-analysis. *The British Journal of Psychiatry*, 221(6), 714–721.

Williams, Z. J., Abdelmessih, P. G., Key, A. P., & Woynaroski, T. G. (2021). Cortical auditory processing of simple stimuli is altered in autism: A meta-analysis of auditory evoked responses. *Biological Psychiatry: Cognitive Neuroscience and Neuroimaging*, 6(8), 767–781.

Wykes, T., Huddy, V., Cellard, C., McGurk, S. R., & Czobor, P. (2011). A meta-analysis of cognitive remediation for schizophrenia: methodology and effect sizes. *American Journal of Psychiatry*, 168(5), 472–485.

Yan, X., Jiang, K., Li, H., Wang, Z., Perkins, K., & Cao, F. (2021). Convergent and divergent brain structural and functional abnormalities associated with developmental dyslexia. *Elife*, 10, e69523.

Zetsche, U., Bürkner, P.-C., & Schulze, L. (2018). Shedding light on the association between repetitive negative thinking and deficits in cognitive control–A meta-analysis. *Clinical psychology review*, 63, 56–65.

2 Andrew's lived experience

Attention-Deficit/Hyperactivity Disorder (ADHD) is a neurodevelopmental condition. While this means that signs are present from childhood, only relatively recent developments in science and medicine have meant that more people are now being recognised as living with ADHD from an early age. This has left many people, currently in adulthood, discovering this part of their identity later in life. The impacts of this are potentially far-reaching. While a diagnosis at any stage in life can bring relief by explaining long-standing difficulties with learning, relationships, and personal organisation and an opportunity to develop new ways of coping to improve quality of life, it can also bring about a sense of frustration, perhaps due to missed opportunities, recognised damage to self-esteem, and possibly challenging emotions such as regret, and shame, felt over past experiences.

Hello, my name is Andrew. I'm a 30-something-year-old male and my entire life I've never really 'fit in'. From the very beginning of my time at school, to the end, I felt like I had no idea what I had to do. In infant and junior school, I would look around the classroom at the other students all focused and getting on with their work only to feel a slight panic as I'd look back at my workbook, clueless. The teachers would frequently tell me not to stare out the window, as I was deep in daydreams. I'm sure that's probably one of the reasons why I had no idea what I had to do.

I had absolutely no interest in reading or anything else that required any prolonged concentration or mental effort. I much preferred fun, immediate gratification. Initially, this was nothing more than looking forward to playtime, trying to extract as much time as possible with friends after school and finding it absolutely impossible to sit down and focus on my homework. But, by year 4, while I was still only eight to nine years old, I became very interested in the opposite sex. I would chase girls around the playground (in an innocent, childlike way, of course), play footsie in class, write them little notes (that occasionally got me in trouble), think about them when I was at home and even try to convince my parents to let me invite some of them over.

DOI: 10.4324/9781032714745-2

As I moved into high school, the daydreaming continued and my fascination with girls grew, and now I also had video games to help distract me. I remember trying to focus and get on with my work countless times (as I did want to do well), but I was so easily distracted and as soon as I was distracted, that was it. It would be a real effort to refocus, only for the next little distraction to knock me off course. I eventually found music and decided that was going to be my love and the thing I was totally obsessed with, so I started playing guitar. With that in mind, you'd expect me to have done well, but the reality is that by this time I had become quite 'rebellious', even to the point I was speaking back to teachers. At 15 I told my school music teacher that I simply didn't care for the confines of the marking scheme because it 'constrained my creativity' and that was why I was struggling. In reality, I just couldn't focus on learning how to read music and how to compose simple pieces for my music studies in secondary school. I just wanted to play what I thought was fun and nothing else.

This attitude permeated throughout my high school career, which, as I approached the end, only revolved around dating (with each relationship ending catastrophically), video games, music, and messing about with my friends. Outside of school, I was talking back to my parents, with whom I felt I had no real connection, and I had complaints from the neighbours for playing music too loud. Maybe this sounds like being a normal teenager, but it resulted in my leaving school with a few Cs, Ds, and even a couple of Es in my GCSEs and delusions of grandeur for my band, yet no one suggested that I might need some extra help or understanding of any kind.

After finishing school, I wanted to follow my friends to college to study music, but my parents simply wouldn't allow it (they felt it wasn't a safe career option). As I didn't want to study anything else, I ended up working as a greenkeeper on a golf course, where I accidentally crashed a highly expensive piece of equipment because I wasn't paying attention. I did this job for about 18 months, while my friends were at college doing their A-Levels, until I realised I too, felt ready to go back to education.

Despite my secondary school grades I managed to get a place at a local college to study three subjects that sounded mildly interesting. Unfortunately, the issues I had at school only worsened when I was re-introduced to a group of people of my own age at college, after 18 months of solitude on a golf course. Once again, I daydreamed, chatted too much, focused on several girls simultaneously and started to drink to excess, smoke, and generally act recklessly. In my first year of college, I had to work part-time to help pay for transport and various other things. I got fired from two jobs because I chose not to attend shifts, had a generally poor attitude, and couldn't work as fast as my colleagues. It wasn't until the end of my final year of college, at the age of 19, nearly 20, where I found an ability to focus and start to enjoy the feeling of being focused and taking things more seriously. However, it was too late. My earlier inhibition meant I left college, four years after school, once again with Cs.

However, I feel this was a wakeup call for me. Somehow, I managed to get a place at an up-and-coming university to study the only subject I was interested in. I nearly failed my first year, but something clicked in the summer between then and my second year and from then until now I have been able to focus better, enjoy the process, and take things much more seriously. I think the big difference was that I finally found something I genuinely loved to learn, and I realised there were ways for me to study, one of them making sure I was totally alone, in my own bedroom or corner of the library with no one else around. This realisation seemed to pay off as I got a 2:1 in my degree and even managed to get a Merit in my master's degree.

However, I notice I still daydream a lot; I struggle with authority figures at work, home, and wherever else they might be, and, in every job I am often not as quick as my colleagues at doing routine tasks. Despite my efforts to do my absolute best, I also take risks, and I still want to do things to get that immediate gratification. Since my late 20s that has come in the form of dating a lot, obsessing over and buying things I don't have money for and don't really need, going to luxury hotels and restaurants, trying to experience fast cars, planning expensive holidays, all while ignoring actual financial obligations I have. I still struggle with being impulsive. To be frank, it is exhausting. I feel I am constantly battling myself and I'm sure much of this is why it is taking me much longer to 'achieve my potential', when compared to my friends.

After reflecting on all of this in my late 20s I realised, for the first time, I genuinely don't think, feel, or act like the majority of people I had met. So, after learning more about mental health and neurodiversity from courses, friends, family, and the internet, I decided to speak to a clinical psychologist about the possibility of various diagnoses and a consultant psychiatrist to explore the possibility of having ADHD. After a formal, three-stage assessment that included both myself and my parents, I was diagnosed with ADHD at the age of 32. At the time of diagnosis, the consultant said: 'Don't worry, it isn't your fault'. Those words will always stick with me. When you find it so hard to focus and understand things and stand in line with everyone else, you often don't excel. You berate yourself for these perceived failures and feel so frustrated by them because you feel you don't have as much control as you'd like. So, hearing a highly experienced specialist say: 'It's not your fault' teaches you that you aren't a bad person or a failure, you're just wired differently (but of course it is important to make the effort to understand yourself, with an aim to improve and be as accountable as possible). That is why I am contributing to this book, to help people understand what it's like to exist with ADHD and how I think things could be improved.

On several occasions I have mentioned the challenges I've had with how I think and feel and relate to others. This is something I want to describe in more detail. I believe one of the most prominent factors for this is emotional dysregulation, associated with ADHD.

I find it very easy to experience intense emotions, whether that be highs or lows. Some mornings I will wake up and feel like one part of my mind is travelling at hyper-speed and the other part is left hopelessly trying to catch up. I also find it hard to try new things and it can take a lot of effort for me to understand and agree with the perspectives and actions of others.

I can get obsessed with things, people, or concepts that I find very alluring. Practically, this means that items, places, or new hobbies will dominate my thoughts to the point I will find it hard to consider much else and find it even harder to get to and stay asleep. I will feel compelled to read and watch everything to do with it. I think this is probably how I was able to succeed when I finally found a subject at university that genuinely interested me. I will research it relentlessly until I either get the small high from buying/doing it or conclude, after significant back and forth, it cannot be done, which is often followed by a low period.

I can also have strong opinions, a strong will, and I am naturally assertive. This can cause issues because I easily assume leadership responsibility, I feel too comfortable sharing my opinions or suggestions when it might not be appropriate, and I often don't recognise or like to follow 'official processes' that involve a long, drawn-out procedure. Despite my efforts, I am also still very impatient and can feel quite anxious if things aren't done, followed up, or finalised quickly, which often means I send a second, very polite, email or text or I decide to fret at home. I think this is compounded by the fact I tend to forget things quite quickly and start thinking about something else even faster. It means if I don't do things then and there, I may never do them at all.

However, I also often lack confidence, which may be partly a consequence of how difficult I find it to focus and succeed. This often means that I need a lot of reassurance, despite being assertive and holding strong opinions, which I'm sure is a challenging route to manoeuvre.

When I was younger, in my late teens and early 20s, I was also very impulsive and jealous. This combination of various emotional states resulted in unnecessarily explosive, impulsive, and intense verbal reactions to things, to my partners, family, friends, and colleagues. If I saw or experienced something I didn't like or I felt was unjust I would react immediately, at 80% intensity. There would be no consideration for feelings, fallout, or the future. Almost immediately after these outbursts I would feel intensely regretful and apologise profusely and internally berate myself. Over the years, I have lost friends and partners because my reactions, over time, were simply too much to accept.

As I mentioned, I am a naturally caring and sensitive person. I do want to do well and be good and treat others as they deserve to be treated. I am always reminding myself of this. Over the years I have become increasingly aware of how challenging I find most things and the effect I have on other people. It has caused significant anxiety and moments of depression.

However, thanks to my recent ADHD diagnosis and my naturally introspective nature, I have been able to find help and learn how to better manage my emotions, reactions, and self-esteem, which has been very effective. I now have more time in exams, I am better equipped to hold myself back from being impulsive, I am more aware of when I want to be reckless or take unnecessary risks and how to restrain myself. I believe I've also benefited from learning more about 'neurodivergence', as it is currently described. Despite my 'differences', I have always been a caring and sensitive person deep down, so to know there is a wider reason for a lot of my complex thoughts, feelings, and behaviours is comforting. I also keep a very well organised digital diary and notebook on my phone. I put everything in it that I need to do, just so I don't forget.

In conclusion, I would like the person reading this to understand that there is more to neurodiversity than the 'naughty child in school' or a simple excuse for being not as quick as colleagues or classmates or for being a 'difficult person'. The experiences of those with neurodiversity are wide ranging and often very challenging. I have had friends, partners, and employers all doubt the credibility and seriousness of ADHD and its diagnosis. Yet, I have spoken to many people with ADHD, and we present in many of the same ways and have faced the same challenges, especially showing intense reactions, thinking differently to others, and lagging behind our counterparts, despite showing the potential for success and trying our hardest. All I ask is that you accept neurodiversity as a real state, that you come to accept people with neurodiversity genuinely struggle and are often at war with themselves, feeling great frustration and lacking self-esteem and if you can, find a way to accommodate for them, whether that be at work, school, home, or in a relationship.

Commentary

Andrew describes how as a child he never really felt that he fit in, whether that was at school or at home. He discusses how his lack of focus affected his ability to not only succeed at school but also how this affected his relationships, whether this be romantically or platonically. It is clear that Andrew spent the majority of his life until his diagnosis, a relatively short while ago, being aware of his difference but not being able to label it. He describes the moment when a doctor gives him the diagnosis of ADHD and explains that his difficulties were not his fault. This then gave Andrew permission to see that the way he viewed the world, with the emotional dysregulation he experiences, was because he is wired differently to most people, not because he is a bad person.

Andrew highlights some of the ways that ADHD has affected his life so far; he explains how it makes it difficult for him to focus on things unless he is very interested in the topic. When this happens, he can then become so invested in this new area that he neglects other aspects of his life. He also describes how he can have very strong opinions at times and can come across as assertive. This can make him a natural leader, but due to his difficulty with focus, he can become frustrated or daydream easily, which can then lead to further difficulties. These have led to multiple problems throughout his life, which in turn have knocked his confidence in his abilities. Andrew also details how his ADHD interferes with his ability to control his emotions, leading to impulsivity or verbal outbursts towards others. This would result in him berating himself and lowering his self-esteem further.

Despite these challenges, Andrew has learnt how to manage his ADHD, with the help of professionals. He has learnt strategies to better handle his emotions, to notice when he is being reckless, and to restrain himself as well as discovering practical solutions for his difficulties with memory and focus.

Andrew closes his chapter by highlighting how being neurodivergent is much more than just having difficulty concentrating. He points out that this is a real state, and that life can be made so much more difficult by being neurodivergent; he asks that others in the neurotypical world accommodate people with neurodivergence.

3   Kanan Tekchandani's lived experience

This chapter explores the experiences of living with an autism spectrum condition or disorder (ASC, also known as ASD). Experiences of this neuro-developmental condition range greatly. Some people with ASC find it harder to reason and problem-solve and may also be diagnosed with an 'intellectual disability', others are very adept at these types of general cognitive skills. However, what defines ASC (with reference to diagnostic guidance) is a difference in (1) social communication and interaction, (2) somewhat restricted and repetitive behaviours, and (3) styles of learning and potentially ways of moving and paying attention. Therefore, a person with ASC could be said to experience challenges with 'higher-level' specific cognitive functions (also known as 'executive functions'). Despite this, there has been a societal push in recent years to not frame ASC as a 'problem' or 'difficulty', especially when someone is very able in terms of general or 'intellectual' cognitive functions. This neuro-affirming position emphasises the unique strengths of people who identify with ASC and places the emphasis on any 'differences' relating to the need for societal shifts.

I agree that the neurodivergent experience is sometimes difficult, but it is the environment and the perception of neurodivergence that often disables us. I like to compare to this saying 'everyone is a genius, but if you take a fish out of water, it will think it is stupid' (anonymous). Now, I do not actually believe everyone is a genius, but rather, we all have our strengths and challenges. It is when we are consciously aware of our challenges, as well as the strategies we have likely subconsciously developed over time to survive and get things done, that we are better able to cope and thrive. Rather than see us as less than, we can see ourselves as struggling in an environment that does not necessarily suit us. When we do this, we can give ourselves self-compassion and a more positive self-perception. We realise how much harder we are having to work simply to exist in society compared to neurotypicals.

The one size fits all, and the push for all to meet minimum criteria troubles me. What if you cannot meet these expectations?

DOI: 10.4324/9781032714745-3

I myself have been lucky enough to have the intelligence, skill, and resilience to develop strategies that have got me through what was once quite debilitating. For example, situational mutism is a big memory from part of my childhood school years. I recall when teachers would ask me a question in class, and the overwhelming feelings that prevented me from saying anything or even moving. I now understand that I was experiencing a threat response, and my reaction was to freeze to stay 'safe' – psychologically, emotionally, and socially.

So, in this case, being unaware of my nervous system and unaware of my sensitive, highly perceptive wiring, I was confused and simply on 'high alert'. My reaction was automatic, simply part of my nervous system responding to the environment I was in. Socially I did not feel safe to be myself or to speak. This incredible fear caused me to shut down in those moments.

Even when with friends I would not be able to speak sometimes – usually when there was more than one or two of them. One good friend, one safe person, I could handle a conversation with; two, became a bit challenging, and three or more, it was hard for my brain to keep up with the conversations and to figure out how I was expected to respond. So, this was often about expectations and not about my natural way of being. I could not access that part yet with other people because I was too busy worrying about and focusing on the external demands and expectations, how to be 'normal' or 'like everyone else'.

Fear was heightened when going into a completely unknown place or situation such as starting a new school, going to a new house, or meeting new people. The more unknowns the harder it was to feel relaxed and to just keep up, let alone enjoy and learn.

Hiding in the toilet at school is another memory that is clear in my mind. The safety of a locked cubicle door, and the downplaying of the sound of multiple loud voices and screams of excitement during breaktime. It was a refuge and time to allow my breathing and my body to start to come down from high alert.

Even now hiding or avoiding is a strategy that I can choose to implement. As an adult, figuring out what was going on in relation to my emotions, physical experiences, my nervous system response, and my thoughts to any particular environment and situation has given me the relief to be fully aware of how all these things are connected. Like a recipe for any particular outcome I choose, I now have a sense of power and control. I get to decide what environment I want and know I can thrive in.

Aspects of the environment I can modify include: 1) the people, 2) the physical space of the room, building, or location, 3) the sound, smells, touch, taste, and the energy of a space, 4) my own physical, emotional, energetic state, i.e., my body, 5) the time and rhythm of life, and 6) the garden of my mind. These I discovered and got to know through my yoga, mindfulness, and meditation practices. In yogic philosophy we have five koshas (layers of being), yet most of the world only really acknowledges and is aware of the gross layer of the body and sometimes the mind.

In the right environment that is somewhere quiet, soothing, minimal, organised, soft, and familiar, then I am able to demonstrate my abilities. Hence working from home turned out to be my saviour and the way that I could truly thrive in life and do meaningful work – that is, being of service to others in a significantly impactful way.

In the right environment, my nervous system no longer needs 'managing' and keeping in a safe window of tolerance. In the right environment, I am not fighting to just exist in the space, but I can actually conserve that energy and channel it into something that is useful for others, that can create a positive impact on humanity.

As a trauma-informed coach for adults, I get to help others readjust their world so that they can step into living out their purpose. Each of us has unique skills and gifts so to me it seems a great shame to waste this just because we have found ourselves in the wrong environment; but what is even worse is not realising that we are in an environment that does not necessarily suit us and may actually be holding us back from living out our potential.

The tragedy is not knowing what you don't know. The quicker we can become self-aware the quicker we can get 'unstuck' in simply trying to survive, the quicker we get to feel useful, purposeful, and own our power.

It begins with us. We need to find sources of information and guidance to help us get on to the right path. With this guidance we can learn how we operate as a whole human being and this self-awareness allows us to self-advocate for our needs. It is extremely unrealistic to just hope that the world will want to, and will actually, understand us. We also have our part to play. It is a collaboration between all humans; that is how we can become a unified force for good.

If we simply rely on others to learn about us and we believe that our lot or the world is unfair, then we give away our power. We no longer have the determination and motivation to make changes for ourselves. Essentially, we allow ourselves to feel disempowered. We lose ourselves to the beliefs of our culture and our society. Sometimes we are not actually helpless, but we have 'learnt helplessness'.

What we expect of ourselves and what others expect of us can vary but is relevant. To be able to identify these demands and be honest, realistic, and non-judgemental about our ability to meet these expectations is important. Yes, others may have expectations of us, but what is our part in this too? What are we allowing or tolerating? How are we bending ourselves to fit, not realising that perhaps we could try a different way and that this different way is a normal and acceptable, healthy variation even?

So, in summary, I do believe we can feel disabled by our environment and yes, we do have limitations on our capacities. However, we can choose to adapt to our environment or to advocate for ourselves; ask for our environment to be changed to suit us in the most helpful way possible. Often these changes not only help us but ripple out and support others too.

If we look at autism, I feel it has been stigmatised, and I believe this is due to the stereotyping that has occurred. It is seen mainly as a disorder, a negative, something to be feared or worried about. It is seen as being less than rather than different. I see it as different – sometimes it holds us back more than it would others and sometimes it helps us more than it would others. So, different. If a neurotypical person was put in an environment that did not suit them, e.g., it was noisy, unstructured, and frightening, then I am sure that this would affect how they think and behave. It is not just people who are or who identify as neurodivergent that get affected by their environment.

The way people see us and then the way they interact with us sends us very clear but subconscious messages. This happens from when we are little all the way through to adulthood. When we are pitied or helped it has the potential to weaken us, to make us believe that we cannot do certain things – learned helplessness, essentially. If we are not treated in this way, but collaborated with, dynamically, so that both parties learn to understand each other better, then we can be flexible and adaptable.

If we take a more balanced view of this, we know that it is indeed a spectrum and so where we fall on that spectrum also impacts how we interact with the environment we exist in. Some people who are or who identify as autistic are non-verbal, while others of us have learnt and mastered the skill of speaking conversationally. Even where there is a possibility to learn a skill, sometimes this comes at a cost for it is not always something that comes naturally to us. It takes effort. It can be exhausting and leave us on the edge of burnout unable to simply function or take care of ourselves.

However, some of us have a great capacity for learning – especially when we get to hyperfocus on our passions. Personally, I have loved allowing myself to follow the rabbit holes of human needs psychology, gifted psychology, spirituality, trauma, and somatic experiences. When I give myself permission to follow my loves it makes my life flow more easily. When I force myself to do something unnatural, to go with the mainstream beliefs of what it means to be successful, academically, or professionally, that is when I lose my natural energy.

I first discovered I was autistic in my 40s after my son was diagnosed. It was not long after that, maybe a couple of years later, that I then discovered the concept of being gifted – or should I say giftedness – and the gifted community found me.

I had never really thought much about the concept of giftedness or its relevance in my or my family's life. Again, the stereotype of a boy genius like Einstein who stuck out like an awkward sore thumb was not something I related to.

If I had known that giftedness is also a spectrum and that it came with great social and emotional challenges, then perhaps I would have discovered it earlier. But it was only when I met and spoke with other gifted and often 'twice exceptional' adults that I felt that resonance that told me: 'I'm like them', 'I belong'. I felt a sense of I'm like them and they're like me.

Being gifted is seen as a 'gift'. This term comes with connotations of being so smart that you are looked up to, respected, successful, and rich.

It is strange that I had not considered all the other aspects in a holistic way until now. Now I can see how disabling it can be to overthink and over complicate every thought and decision you have to make in life. Where the complex ability of the brain actually can slow you down or prevent you from finding a place to settle. Difficulty in resolving and completing tasks or answering questions because there is always another perspective or possibility to consider.

Overwhelm and procrastination, constant self-questioning, and rethinking takes up a lot of time and energy. That's why I have a poster which says 'Keep It Simple' – it is a reminder that although my mind can do great things, sometimes my brain takes over and becomes unhelpful. It tells me that I need to identify when the right time is for my brain to go into overdrive and also to identify when it is time to stop relying on the brain and drop my connection into my body and intuition.

When you have this self-awareness and the ability to adapt yourself or the situation to work for you then that makes you extremely resourceful and able to cope with many of the demands that we all get thrown at us. Innovation and healthy adaptation are so important. It is both the person as well as the environment (thoughts, body feelings, nervous system, other people and how we interrelate) coming together to result in an outcome. Change and educate any of these factors and you get a different outcome. Do it with clear intention to understand the parts of each system as well as how the system interacts, then we get to create a more harmonious and effective human society.

Commentary

This chapter emphasises how many people with the diagnosis of autism feel that they are different from an early age, regardless of whether they have been given a formal diagnosis or not. The author did not, in fact, discover that they were autistic until they were in their 40s, however they were able to highlight several aspects of their lives where they knew they were neuroatypical. This included periods of high stress and fear as they grew up. The author described how the discovery of what they needed their environment to be like for them to succeed, was a game changer. They stress how important it is not to see being different as a problem, but rather to see it as an environmental issue; if the environment is not suited for an individual, how can they be expected to thrive, independent of whether they are neurodivergent or not? They describe how many neurodivergent people, in their experience, allow themselves to become disempowered which leads to learned helplessness. The author speaks about their experience of discovering the 'gifted' community and how it felt to find others who were like them.

It is important that people are given the chance and taught the skills they need to build the self-awareness to create and adapt their situations to work for them. In doing this, people with neurodiversity will increase their resourcefulness and, in the author's words, they will be able to 'cope with many of the demands that get thrown at us'. As highlighted in the introductory chapter, perceiving and interacting with the world differently from societal norms does not prevent individuals from living well, however, these individuals may need additional support, especially with their environments, to flourish in all aspects of life.

4   Janice's lived experience

Multiple Sclerosis (MS) is a chronic health condition that damages the nerves in the brain and spinal cord. It is autoimmune in origin, meaning that the body's immune system attacks the fatty insulating layer that surrounds nerve cells, 'myelin', that protects the nerves and allows electrical impulses to travel quickly and efficiently. This damage disrupts nerve function and can cause a range of symptoms that can include cognitive and visual function to movement-related difficulties but are different for everyone. The first time a person experiences symptoms that look like MS, they can be diagnosed with Clinically Isolated Syndrome (CIS). Diagnosing MS is based on the 'McDonald criteria', which look for evidence of damage to the central nervous system (CNS) and must be spread out across time and space (i.e., within different regions of the CNS). Sadly, there is no cure for MS. Treatment focusses on helping to manage symptoms and slowing the progression of damage to the CNS, with the most advanced being a stem cell transplant, a procedure that uses chemotherapy to remove harmful immune cells and then replace them with the person's own stem cells. This process is called autologous hematopoietic stem cell transplantation (aHSCT).

My name is Janice, and I was diagnosed with a Clinically Isolated Syndrome (CIS) in 2008 and formally diagnosed with multiple sclerosis (MS) in 2010. In this chapter, I wish to share some of my experiences in coping with this condition and highlight the various ways in which brain structure and/or function can be affected. Although people with MS may have a 'different from average' structure and/or function to their brain or nervous system due to a process called 'demyelination', each person with the condition is also very different, both physically and psychologically. It is often underappreciated how diverse the symptoms of MS are. And, as part of this, it is possible that other types of 'neurodivergence' that preceded the diagnosis of MS, such as neurodevelopmental conditions, are also present. Speaking from personal experience, my father was almost certainly undiagnosed as on the autistic spectrum, and it has been suggested that, like so many people, I

DOI: 10.4324/9781032714745-4

am also at the 'mild' end of this spectrum but sometimes the condition commonly goes unrecognised in females.

When it comes to my brain, I recall being told early in my diagnosis of MS that I had 'significant brain shrinkage for my age'. It is not unusual for MS to affect the structure of the brain, and the impact can be far-ranging, resulting in physical, cognitive, and emotional changes; for example, some people find it hard to control their emotions, leading to crying or laughing at times that may feel unusual or unprompted. The intensity of these symptoms can vary over time. I certainly experienced my own psychological impact of having such an unpredictable condition. One day I was going out running, skiing; playing golf, and working in my own physiotherapy practice; the next I was climbing the stairs on my hands and knees and having to be pushed in a wheelchair when I left the house. I couldn't even propel myself in a wheelchair because my right arm would fatigue rapidly. Obviously, this had a massive psychological impact. I didn't have a car, and my friends were all working, and it wasn't too long before depression ensued. It is really common for people with MS to be diagnosed with anxiety and depression, which can also have an impact on brain function. Loneliness can be a large contributing factor to depression. As a person becomes more disabled it is not uncommon for marriages to break down. I have been luckier than most that my husband has stood by and supported me. I found that when I was no longer able to work I had considerably less social interaction. Some social activities are no longer possible as people become increasingly disabled, prior to MS much of my social life revolved around playing golf. Even meeting a friend for a coffee or lunch can be more difficult when disabled, requiring easy parking nearby and disabled access to venues. This isn't always taken into consideration when non-disabled friends organise a get together for a group.

I recall that when I struggled with anxiety I suffered with rumination and worry, but this is something that I experienced prior to MS. A combination of Cognitive Behavioural Therapy (CBT) and being in remission from MS for over five years (following autologous Haematopoietic Stem Cell Transplant (aHSCT)) has helped significantly with the anxiety.

Medication can also change the way the brain functions and could be seen to contribute to 'neurodiversity'. For example, high dose steroids are frequently prescribed for MS relapses, and it is not uncommon for them to cause dramatic mood swings. I remember when I came off a course experiencing sudden extremes of anger and verbal aggression rapidly followed by euphoria. It felt terrible to experience, and it was very challenging for my husband to deal with.

Taking MS, neurodevelopmental processes, psychological responses, and medication all into account, I identify with being 'neurodivergent'. This was most obvious to me when I experienced a relapse of MS. When the MS was aggressive, I suffered with anxiety and depression, I felt as if I had 'lost me'. Throughout my life I have used exercise for stress relief and suddenly this release was lost to me. I was very aware that before HSCT I had experienced significant cognitive problems. My attention span was poor, I had been unable to read a book or watch a film for years. I previously experienced 'word finding difficulties' with everyday words, e.g., Prosecco, yet I could remember medical terms.

Prior to aHSCT I had many unmet needs. I think that it is important that people realise the effect that MS has on cognition, mood, and behaviour. I think that MS is generally seen as a condition that causes physical disability, but the other effects are less well understood. I have spoken to people who have had issues in the workplace where this has not been taken into consideration. One particular man with MS was considered by his colleagues as having become moody over time but he received no empathy from most of them because they had not realised that it was through no fault of his own. I am not sure that these changes are always understood by people who are newly diagnosed and this lack of understanding about what is happening to their mood may contribute to the relatively high suicide rate of those in the early stages of the disease. Despite steroids being prescribed for MS relapses, patients are often not warned about steroid psychosis and the dramatic impact that this has on mood and behaviour.

I believe that the greatest unmet need is that of monitoring cognitive decline in people with MS. There is no baseline measure of this so there is nothing to compare it to. In 2013 I reported that I was having some memory problems. An occupational therapist came to my home and performed a memory test. She told me that I had done extremely well but added that the test wasn't very sensitive because it had been designed for people with dementia!

The chemotherapy used for aHSCT is known to cause cognitive difficulties that may last for up to a year afterward. I don't know how some patients returned to professional jobs after three months, but I know that some employers gave them no choice. In recent years neuroplasticity has gradually led to a significant recovery with this. I have now been in remission five years and feel like myself again, possibly still neurodivergent in the sense of being mildly on the autistic spectrum still but the anger and frustration have gone because I am able again.

Physically, at present, I am one of the fortunate ones that have extremely mild symptoms (non-symptomatic bladder retention and mild weakness in the right gluteal muscles that would be undetected by a neurologist). My memory is still not as good; I suffer with 'brain blanks' regarding significant events that have happened in my life. As I type this I still can't always think of the word or phrase that I want to use and often question whether I have used a word correctly in a certain context. I know that my vocabulary is a great deal more restricted than it was pre-MS. Although I have made a relatively good recovery (through Neuroplasticity) and I am on no treatment for MS, I do recognise that aHSCT for MS is not permanent for everyone; it is possible that I could return to having more difficulties one day. Physically I do not believe that I could concentrate for long enough to do the updating qualifications to return to pursuing a physiotherapy career, so I have had to come to terms with this change in my life. Despite having mild weakness in my right gluteals, which makes me more susceptible to sports injuries, I play golf at club competition level; regularly attend exercise classes and go to the gym. My neurologist recently downgraded my EDSS (Expanded Disability Status Scale) to two, which is very low.

I am very aware that I have much fewer unmet needs currently compared to the majority of people with MS because I was able to access aHSCT privately abroad and I have been extremely fortunate to make a degree of improvement both cognitively and physically and as discussed my physical health subsequently has a major impact on emotional health.

I no longer qualify for a Blue Badge nor any benefits and therefore do not consider myself as a person with a disability. I am able to do so much more than I was able to do and prefer to focus on this. I am physically very active again; can concentrate on watching something on television for two hours and can read a book for pleasure again. I decided from the beginning that I wanted to be identified as 'Janice', 'the lady who used to be a physio'; 'the lady who was friendly and tried to help people'; 'the lady that I went to school with, who was annoying because she always won the sports awards and got good grades in exams'. I wanted to be known as anything but 'the lady with MS'!

Commentary

Janice is not the 'lady with MS', rather she is the lady with many other important factors in her life, who just happens to also have been given the diagnosis of MS. In her chapter Janice highlights how MS has a detrimental effect on all aspects of a person's life, whether this be physical, cognitive, or emotional. In her experience, before she had the restorative procedure, the MS was impacting all of these areas. She was having more difficulty moving around, had to give up her job as a physiotherapist, her social life had reduced considerably, and she was noticing more cognitive deficits, such as word finding difficulties and memory impairments. With this in mind, she does identify as neurodivergent and highlights that she is probably mildly autistic as well as having a diagnosis of MS. Janice goes on to say that she is aware that she is very lucky to have had the aHSCT procedure as this has resulted in most of her impairments improving: she no longer qualifies for any benefits or for a blue badge rather she is back to her old self, being active, reading, and watching films, things that were a struggle before the aHSCT.

5   Sam Shephard's lived experience

Traumatic brain injuries (TBI) are a specific type of acquired brain injury (ABI) that can occur when external forces cause the brain to move rapidly within the skull, leading to various forms of damage to the brain, commonly seen as a result of accidents such as road traffic accidents (RTA). This damage can cause permanent change to the way that the brain is structured and / or functions. Secondary complications due to infection, for example, and additional chronic neurological health conditions, such as epilepsy, can result. People can be left with a combination of physical, cognitive, and emotional issues. Many people have to work out ways of coping (also known as compensatory strategies) to manage in daily life. Some people find that they do more than just cope, though, they succeed.

I had a severe traumatic brain injury (TBI) when I was 21, on 14 April 1996. As a result of either infections or surgeries, I have subsequently experienced additional brain injuries. After one such surgery, I acquired epilepsy. My brain has changed, over these last 28 years, and with these changes, I have learnt to use it differently. At times I have wondered whether I am still the person I was on 13 April 1996. Whilst I'm now satisfied that the answer to this question is yes, I certainly experience and interact with the world very differently now to how I did by the end of my first two decades of life.

For much of the time since the accident, I didn't understand how I had changed, nor how to overcome the many barriers I faced to moving forwards in life. I didn't even know they existed, never mind what they were – I just felt stuck, lost. Although feeling apart from much of society – isolated and different – I couldn't explain quite how this was, and I had no one-else in my life with an 'acquired brain injury' (ABI). I had become neuro-atypical but did not associate with this label until many years later, following a chance conversation in 2017 with someone who told me they were neurodiverse. Discussing my experience of learning (I had returned to education), they too considered me neurodiverse. This felt novel to me, at first not fitting with my developing identity, or the ABI community I had finally begun to feel a part of. Yet with

DOI: 10.4324/9781032714745-5

knowledge, I began to see many of my experiences, some of them challenging, were shared with some people with neurodiverse conditions. My challenges and experiences were acquired rather than developmental.

The two most significant effects of my ABI are degraded memory, and persistent fatigue, together forming ongoing, inter-related, challenges. With further injury, my memory has altered – I am already experiencing worsened short-term and working memory, I've seen noticeable changes in my spatial and visual memory following specific surgeries. I attempt to manage my fatigue in a way that doesn't impact my cognition, mood, or social engagement – although I now feel successful at this, just ask my family what I'm like when I don't get this right!

Early on, I tried to mask the fact that I didn't recognise people I'd already met, or that I'd forgotten things. This made me feel comfortable but put me at risk of people calling my bluff, of revealing that I was not as in control as I appeared. Really, I was being dishonest with myself. I now know to tell people I can have a problem with faces and names from the outset, avoiding this kind of in-the-moment stress. New environments with lots of people can be reliable triggers for stress and overload. When these social and environmental stressors begin to take their toll on me, I know to take a break from my surroundings, even for a short while, otherwise I become overloaded and cannot be an engaged (or engaging) part of the world around me.

I also experience cognitive difficulties in the form of déjà vu, a type of epileptic seizure, having acquired temporal lobe epilepsy. Now largely controlled by medication, this can still be disorientating as I try to establish whether something has happened before, or whether it may be similar to something else. It took me a long time to recognise seizures for what they are, and despite the medication, they are especially likely when I'm in new places. I try to quell the urge to tell myself 'Sure, I know this place, it's right over here…' – without consciously putting these brakes on I'm very likely to end up lost in a place I thought I knew.

Further injuries have increased my fatigue – something I experience it as more than just physical tiredness; it is also a cognitive, emotional, social, and environmental experience, that is both variable and fluctuating. This has been disabling, yet whilst still inconvenient, I can navigate life with adjustments, adaptation, and anticipation of what might be difficult. This latter point has been born from experiential knowledge. Fortunately, I have arrived at a place where I can see these cognitive differences as changed ways of living rather than disabilities. I may be lucky in this though, given the social factors people with brain injury are confronted with. This includes stigma – an indicator of lack of knowledge and is an experience common to many people in the neurodiverse community, regardless of its definition.

As areas of my brain's functioning, and its effects on my daily life, changed with time, my acceptance of the need to use support strategies improved. This helped me regain and maintain a stronger degree of control over the sometimes challenging influences on my thoughts and feelings of my environment, the people within it, and the passage of time. However, this acceptance didn't come naturally; at first, I didn't see the need to use diaries, calendars, lists, etcetera, yet with time, I realised I had no choice. Starting out as a person with ABI all those years ago, I didn't use any reliable ways to support my cognition, and like many people, I hoped an organic recovery would quickly return me to the person I had been pre-accident.

With time, experience, and further injuries, I track events and important details, knowing that without this, I risk forgetting both what has happened, what lies ahead, and my relationship to these threads of time. If I don't do this, it is not just my memory that suffers but all the processes it feeds into, including decision-making, planning, and problem-solving.

In the very early 2010s, my support worker was an early adopter of the smartphone, espousing all it could do. With alarms, timers, and notepads, it had all the things I was being encouraged to use, and yet I found the phone confusing and unnecessary, instead sticking with my traditional paper notebooks. Naturally laggard-ish – irrespective of brain injuries – I finally started using a phone to support my cognition. Its use has been an evolutionary matter, but its influence has been revolutionary. I couldn't function anywhere near so well without one. While the smartphone has become so ubiquitous in modern life, why should my use be in any way uncommon? I think this may be because I see it as a cognitive extension of myself, as an outsourcing of capacity for all the things I could no longer reliably hold in my brain post-ABI. I consider this adaptation vitally important, realising during rehab that no matter how much cognitive therapy I did, 'total recall' was not going to happen, nor was it within human grasp. Later, use of technology gave me reassurance that the limitations of my memory could be vastly enhanced, enabling me to engage with the here-and-now, free of the burdens of conscious memory. I was able to relax and become a (slightly) more fun person to be around, someone who was less self-absorbed.

I would like this chapter to contribute to a growing understanding of human difference; something to which the neurodiverse community has made huge inroads in many areas, including media portrayal and employment support. I passionately feel that ABI belongs in this range of human difference – going beyond disability, those of us lucky enough to be able to live well with its effects, are often significantly changed, different to before, yet appearing unchanged to many other people. This is in the wider interest, including people who already fit within the accepted definition of neurodiversity, as anyone can experience ABI. Acceptance of a continuum between disability and difference should empower all people, whether neurodiverse or not, and legitimise support.

An inclusive society would provide improved signposting and clarity – whether this be visibility of services and their purpose, greater understanding of rights and responsibilities, or quite literally, better signposts, so I don't get so easily lost! This shows just how relevant the social model of disability is to cognitive difference – it is often not our difference that disables us, but the systems and attitudes of others. The currently stalled work (as of the eve of the UK 2024 general election) towards a cross-departmental Brain Injury Strategy, modelled on the Autism Act, demonstrates the wide-ranging determinants of health affecting both people with ABI and neurodiversity as it is currently defined. Many of these determinants are the same, triggered by the same causes – most notably a lack of systems-based understanding and compassion (think DWP (Department for Work and Pensions)), and an inflexibility to accommodate the needs and aspirations of cognitively different people.

I now work with other people with brain injury, making me acutely aware of just how common issues with memory and executive skills are following ABI, and the effects this can have on identity, function, relationships, and social participation. I may have written about my experience, but these changes affect all of us with ABI – we are all neuro-atypical. 'Neurodiversity' is a social construct – by excluding people with ABI from its protective 'stronger together' umbrella, there is a risk this exclusion becomes a further disabler for this already marginalised group.

Commentary

Sam describes how at age 21 he had a severe TBI, subsequently followed by further brain injuries, which have left him with a number of cognitive deficits and epilepsy. He details how, initially, he was unaware of the barriers that existed for him moving forward in life. Instead, he describes feeling isolated and lost, with no one else in his life with an ABI to compare life expectations with. A chance conversation many years after the first brain injury led him to the realisation that he was now neuroatypical. This led him to see how many of the challenges he had lived with since his brain injuries, are also shared by many people who identify as neurodiverse – Sam highlights the main difference being that his differences are acquired rather than developmental.

Sam explains to the reader how his brain injury has affected him over the years, including poor memory, déjà vu seizures and fatigue. He has a positive outlook on these cognitive differences, choosing to view them as changed ways of living rather than disabilities. Following this, Sam reflects on how as time and his brain changed, so did his acceptance of needing to learn strategies to manage his cognitive abilities. He acknowledged that this was not an easy thing to accept, but slowly he started using pen-and-paper notebooks; as time and technology have progressed, he now uses his mobile phone as a supplement for his brain, describing his smartphone as 'a cognitive extension of myself'.

Sam finishes his chapter by highlighting the need for ABI not to be seen as separate from neurodiversity, but rather for it to be encompassed by the umbrella term of neurodivergence. He hopes for a more inclusive society and points out that by excluding people with ABI from the 'stronger together' neurodiverse population, then we risk adding to the marginalisation of this already disempowered group.

6 Katy's lived experience

Pituitary adenomas, also known as pituitary neuroendocrine tumours, are noncancerous tumours that develop in the pituitary gland, in the brain. Common signs include severe visual problems due to being close to, pressing on, the optic nerve, and surgery can be a necessary treatment. This tumour and subsequent surgery can lead to long-term cognitive changes as well as hormone imbalances. However, relatively little is known about this type of acquired brain injury (ABI), leading to common misunderstandings and experiences of coping with the effects in silence. This chapter also raises the important point that acquired neurodiversity may not be experienced in isolation of neurodevelopmental conditions seen as neurodiverse. We must remember to consider a holistic view of each person across their lifespan.

Katy and I[1] sat together quietly for a minute or so, while she gathered her thoughts on why the topic of neurodiversity was important to her. She said, 'that was the last question from the list you gave me; that was the toughest one!' She went on to say, 'well, it's kind of a new term,' 'not a lot of people are aware of it.' Katy disclosed that she understood what neurodiversity was, though, as she said she experiences it: 'it's blocking me from succeeding, it's causing me troubles and I feel there is no kind of understanding from other people. Because they see you as a person speaking, functioning, and then suddenly things go sour, and the reaction is that it's all my fault.' Katy is a very gentle and unassuming person, who does not hide from her responsibilities, but sits in front of me with an expression of incredulous guilt. She says, 'I say something rude. Can I help it? No, but they don't understand that. People have the expectation that you are "normal", that you do normal things, and when you step out of that line, there is no acceptance.' But Katy catches herself at this point and revises her words: 'I'm not sure if I'm asking for acceptance, but some kind of understanding that behaviour can be caused by the brain working differently, and it's not purposeful.' Katy looks sad at this point and says: 'Because in those situations, I feel really low. I don't want it to happen, but it still happens, and there's not much I can do

DOI: 10.4324/9781032714745-6

about it because that's my brain and I cannot change the way the brain functions. It is what it is'.

I should say at this point that Katy identifies as neurodiverse for two reasons. First, she identifies as having an undiagnosed neurodevelopmental condition, attention deficit hyperactivity disorder (ADHD), and second, because she was diagnosed with a brain tumour that was operated on and resected several years ago. Talking about her experiences in childhood, Katy told me: 'I kind of learned to live with it [ADHD]. Back then I think I found some kind of coping mechanisms and kind of adapted myself.' She described how she had to learn certain rules like: 'when you need to be quiet, you're quiet. So I forced myself to do that'. She said: 'I realised it when I was in year three. We had a very strict teacher, and I was always in trouble. I was sitting at home, and I was thinking, why am I always in trouble? Why am I different? What is wrong with me? I felt I am different to others. I'm more boisterous, I talk more, I'm louder, I am wilder, I have to control that.' And she describes being able to do that until 'I got my brain tumour, and had surgery' and 'the problems that came up with ADHD spiked up again.' She told me how now she really cannot control her behaviour at times.

I should say at this point, that my impression of Katy is that she is incredibly thoughtful and polite, but that got me wondering, how much under the surface was I not seeing in my conversation with her. I began to think, perhaps I am just falling into the trap that she told me so many others do, of not understanding, even when trying very hard to. When I posed the question to Katy, 'what about other people adapting to your needs?', she looked me straight in the eyes and stated very clearly 'I don't see that happening'. She went on to expand on this: 'It's quite a big ask for anyone to try to adapt to the needs of a neurodiverse person. It's easy for them to forget because everything from the outside looks okay. They don't know what's going on inside me, they can't see it, it's an invisible disability.'

I was curious, yet careful, to ask more about Katy's use of the term disability and how she identifies with this. She told me: 'I just recently started using that word or thinking in this way, because I realise that it's something that disables me. It affects me in great depth. If it was just a mild effect on my life, I wouldn't call it a disability, but this is something that is affecting my life in great detail.' When I asked if she thought she had always had a disability, Katy responded clearly: 'No.' She went on to say: 'I feel that I was neurodiverse, but then after the brain surgery, I feel that I have a disability.' She explained that it was something to do with the age at which she acquired further differences. 'People think that because you're an adult you should know, you should have the life experience to understand how things go, and when you go out of line, people are hard on you, because they cannot imagine that someone would do that by accident.'

I really felt for Katy as she described the lack of control she sometimes feels. She explained: 'I don't want to do something out of the line of normality, but I do it and then I take it as a big knock on me. Like I'm cut to the heart.' She knows she is hard on herself. She knows that it's when she has to give immediate answers when something 'rude' might slip out, but finding the self-compassion and compassion from others to allow her time is hard. 'My brain doesn't function extremely fast; my thinking speed is low'.

Alongside the differences that Kate had already discussed with me, she went on to explain that she has memory issues too. She said that in some situations 'the words don't come to my mind when I just need them. Recalling information is not there. They come back to my mind ten minutes to half an hour later, when I don't need them.' Katy smiles, a bit sadly, as she explains to me that she often fills the gaps with the word 'thing', for example, 'have you seen that thing?' Although she cannot always recall the exact word that she wants to in the moment, she finds it easier to describe the word. In this way, she has been able to adapt to her acquired differences. 'It's my coping mechanism. When I say the "thing", and someone asks what is the "thing"? I will try and think what I use it for.' Sometimes, she says, this strategy even helps her to remember the actual name of the 'thing'. I reflected on what Katy was telling me in this moment and realised that I could empathise with this experience a bit, but also, I recalled other people in my life using the word 'thing' to substitute a word, and it made me think about ways I can help myself and others in these moments. Katy's creativity and determination to work round her difficulties is something we can all learn from, and this filled me with admiration for her.

My respect for Katy grew as she began to describe what she does for a living. 'I work as a teaching assistant in a secondary school and my job is to work with neurodiverse people and those people who cannot access the curriculum'. I asked her, perhaps rather naively, whether she identified with the differences the children she teaches show and she wisely responded: 'Everyone has their own different ways of working and learning'. She told me that 'autism', for example, 'is a spectrum and somebody can be very high functioning and somebody else can have a very high demand of care'. She did draw one parallel to her own experiences, though, when she spoke about how the needs of people with autism are 'very little understood', reminding me of how she began, describing how she felt other people do not always understand her needs. She expanded on her earlier points by saying: 'I don't think anyone can understand it unless they have it. It's internal. Which I call a battle, and it's sometimes very tiring.' But this doesn't stop Kate from using her creativity in every part of her life and channelling her empathy to help others. She smiled, with genuine happiness this time, as she told me: 'I like helping the students with maths and I can be creative about how to teach it. I break it down for them to do the basics.'

At the end of my interview with Katy, I asked her if she could reflect on a key message for someone reading her chapter. She said: 'I would like to say that every day is different. Every day has its own ups and downs. One day or one week can be really down for me and then two weeks later it's really bright, sunny, and everything is fine, and my disability doesn't affect me at all, or it affects me but I'm in the mood that I can cope with it.' I took from this hope; hope that Kate knows that she needs to hold on in there and ride the waves of life.

Commentary

Katy is an extraordinary woman, who has survived a brain tumour and now works as a teaching assistant with students who have learning support needs of their own. She describes how as a young child she became aware that she was different from the rest of her class, being more boisterous and energetic than everyone else. This led to her learning strategies to cope with, what she later discovered, were ADHD traits. She reminisces how she had learnt to manage her life well until she was diagnosed with a brain tumour and then underwent brain surgery. This operation appears to have turned her life upside down: she describes how she will often make rude comments, especially if pushed to answer before she can think through the question fully. She is appalled with herself afterwards, as she is aware that others do not understand her situation – to the external world she looks perfectly 'normal', so why is her behaviour so abnormal? She highlights how when this happens she will feel low as she did not want to respond inappropriately, but her brain does it anyway.

Katy acknowledged that she is both neurodiverse, from ADHD and the consequences of the brain tumour, but also identifies as being disabled. She stated that before the operation, she would not have classed herself as disabled, but now she realises that her condition impacts her life so much that she is disabled. Not only does she have to deal with the fact that she will often say rude or inappropriate things, but she also stated that her processing speed is slow, she has memory difficulties and word-finding issues. Nonetheless, Kate describes creative ways she has learned to manage her difficulties, so much so that they do not interfere with her career.

Overall, Katy's chapter is one of inspiration; she has not allowed her neurodiversity or her acquired disability to get in the way of her becoming a teaching assistant. If anything, having these personal insights makes her a better, more compassionate teacher to the students who are struggling to learn.

Note

1 Katy was interviewed by Sara Simblett.

7 Patrick Litani's lived experience

Definition attention-deficit hyperactivity disorder (ADHD) is characterised by a pattern of inattention, hyperactivity, and impulsivity. Some people experience inattention more and others hyperactivity and impulsively more, others experience all these signs. What is spoken about less is that people with ADHD can demonstrate a heightened sense of creativity or 'divergent thinking', often generating a large number of highly original and, possibly unconventional, ideas. This cognitive difference can be a skill that allows them to excel. However, people with ADHD may struggle with situations that have rules or boundaries, and require a lot of freedom, leaving them with the requirement to seek alternatives to study and direct employment.

Patrick's opening statement to me[1] was 'I am an entrepreneur, and I believe strongly in neurodiversity.' He went on to say, 'I myself have ADHD and I believe that a common understanding of neurodiversity within society is not prevalent'. His strong conviction and passion on this subject was very clear. Patrick explained that he had realised when he was about the age of five that he thought 'differently to others.' He said that he was 'constantly in trouble at school' for 'outbursts of hyperactivity,' and 'couldn't follow any of the instructions given'. He reflected that 'everyone else seemed to be able to sort of fit in or follow the instructions given, but I just seemed not to have that ability at all regardless of knowing that the punishment may be there at the end.' I could sense a deep sadness in the way Patrick spoke about his early life experiences, even more so when he told me 'I felt that I was misunderstood, that my intentions were always of good faith but that my behaviours or things I would say were misconceived and packaged as being a sort of naughty or bad boy.' He explained that this has stayed with him throughout his life, into adulthood, but that now he can see a different perspective too.

Patrick explained that now 'I understand neurodiversity to be quite a large spectrum of differences in brain activity or ways of thinking. I have come to understand it as being people who may be ADHD, autistic, have had a brain injury, or anything that may have affected the way a person's brain

DOI: 10.4324/9781032714745-7

operates.' When I asked him what he thought of a definition of neurodiversity including anyone who, at times, experiences cognitive differences in memory, concentration, planning, problem solving, or decision making, he replied 'I think, hands down, nobody wants to put their hand up and say they've got a disability of any sort, or that they can't focus on particular tasks. I don't think that having any particular diagnosis is desirable, but it is definitely an absolute relief! I think that anyone that classifies themselves as part of the neurodiverse community are not only standing up for themselves but also standing up for all the collective individuals who are experiencing challenges, whether that be the same label or description. The fact that there can be a common understanding is a great way for us to be able to broach the topic.' I listened with great interest as he went on to say, 'a neurodiverse community naturally is in a position to have a stronger voice.'

Patrick did, however, share some words of caution: 'I fear that in popular culture, terms like ADHD are becoming sort of "trendy", but this diagnosis should represent everyday struggles that are real.' This prompted me to ask Patrick more about the things he struggles with himself. He was very happy to explain, saying, 'if I'm interested in something, then I could probably give it my undivided attention. I could probably go on for hours, to the point where I could even avoid sleeping because of how interested and engaged I am with whatever particular subject I'm researching. However, if, for example, I'm set a task, or I'm asked to do something that I don't like doing, then unfortunately, it becomes a sort of fighting battle within myself to even gain the courage to do the first step.' He told me, 'it is actually quite debilitating, people will look at you and think, you are purposefully deciding not to do something, and what I would like to highlight, is what they can't see is, that in the executive function of the brain there is a breakdown where non pleasurable things unfortunately cause anxiety or reluctance to do those activities.' Patrick, again, looked sad at this point in the interview as he explained that he spends a lot of time trying to find coping strategies and mechanisms, inventive ways to mask his inabilities. He gave an example, 'I use my vocal abilities to mask that I am not so good at writing.' He went on to tell me 'I definitely think that ADHD can cause certain problems with mood. Certain difficulties have made me more likely to be sometimes depressed.' Despite the sadness that he sometimes feels, Patrick is able to see a more positive side to his experiences, 'I remember being told once, that you love looking for shortcuts, but actually had I not looked at shortcuts, I probably wouldn't have achieved half the things that I have done in the space of time that I have had in my life so far. So, maybe shortcuts aren't such a bad thing, after all...' He even went on to say, 'when I'm involved in a project that excites me, then, you know, I'm over the moon and I'm happy.' In his job he explained 'I try to look at markets, and ways in which things are operating can be changed. Then, I like to instigate change. I like to look at things from a completely different perspective and sort of make things better. I've been self-employed since the age of 19 because corporate environments were very rigid and didn't allow me to express my creativity.'

We turned at this point in the interview to what Patrick thinks that other people or society in general can do to help people who have neurodiverse needs. He shared with me that 'I think just a level of empathy and understanding. If you know that somebody is not engaging, perhaps rather than making the automatic decision that the person is doing it on purpose, perhaps ask the question, "why"? What I'm advocating for is that there's more onus on society to be aware and supportive of those who have cognitive difficulties.'

Commentary

Patrick is very open about how having ADHD has impacted his life. He is able to not only highlight times when this diagnosis has impeded his progress, such as at school, but he is also able to recognise when having ADHD has helped him, such as in his career. He describes how he can now see an alternative perspective to the 'naughty boy' persona he was labelled with as a child. This is very impressive, as all too often what people are labelled as in childhood is how they will continue to see themselves well into adulthood, if it ever changes at all.

Patrick goes on to say how he sees the term neurodiversity encompassing several different diagnoses, not just autism and ADHD. He acknowledges how no one wants to be different, but that having a diagnosis can be a relief, especially one assumes if the individual already knows that they are different. To be able to name that difference can be empowering to some. However, Patrick also cautions how in popular culture it has become trendy in certain circles to claim that you have ADHD. He highlights that this diagnosis needs to be accompanied by real life everyday struggles.

Patrick ends his interview by emphasising the need for empathy and understanding when working with people who identify as neurodivergent; he believes, rightly, that there needs to be more onus on society to be aware and to support those with cognitive differences.

Note

1 Patrick was interviewed by Sara Simblett.

8 Sara's lived experience

One of the most misunderstood types of neurodiversity included in this book is the impact of mental ill-health on the brain. Some research concludes that psychosis (including affective psychosis experienced as part of bipolar affective disorder) may be the adult endpoint of a declining cognitive trajectory starting in childhood. This means that a person diagnosed with a neurodevelopmental condition may have certain vulnerability factors for later complications affecting mental health, and brain functioning. This evolving path may be confusing and frightening, leaving people feeling very low in confidence. However, if treated similarly to other neurodiverse conditions, people can learn ways to work around their difficulties and live full lives.

The path to writing this chapter has been a journey that has included winding lanes, crossroads, and mountains. Do I see myself as neurodiverse? Yes, I do now. Did I always see myself as neurodiverse? No. To me, this complexity in the way I identify has been influenced by time: time in the sense of time in my life, and time in terms of the time in society.

When I was seven years old my school teacher, a very kind, caring and compassionate person, expressed a concern to my parents that I was 'spending a lot of time under the classroom table…'. With great surprise my parents responded immediately to her concern and followed this teacher's advice to book an appointment with an educational psychologist. I still remember the conversation with my mum after this meeting: 'she rummaged around in my brain'! I now know, as a Clinical Psychologist specialising in Neuropsychology that what she had performed with me was known as a 'cognitive assessment'. She measured my IQ and concluded that I had a problem with something called 'working memory', which I now explain to my clients as 'the ability to hold in mind and use information to complete a task'. Then she gave me the label of 'dyslexia'. At this time in the world, the term 'neurodiversity' did not exist.

Things at school changed a bit after this diagnosis. As a student at an independent school, I was lucky to be in relatively small classes where work

DOI: 10.4324/9781032714745-8

was tailored to my needs. I also saw the most wonderful private tutor who taught me all the 'mental tricks' or 'compensatory strategies' that I still use to this day. Who knew that I'd forever recite the phrase 'Big Elephants And Unusual Tigers – iful' every time I wanted to spell 'beautiful'; or that I would never grasp the concept and be able to use the terms 'left' and 'right' without conjuring up a mental image of a place I'd learnt to ride horses, visualising which way we used to turn on that start of a hack down one of my local country lanes. In fact, I got so good at these compensatory strategies that I went on to achieve more than I ever could have dreamed.

Now I didn't say that there weren't any struggles along the way but for the most part, I was sailing down a straight road, not letting any 'cognitive deficit' get in my way. I still needed support with spelling and grammar at times, I was granted extra time in exams, and I developed a huge dependency on a paper diary! But at some point, I stopped using the label of 'dyslexia' to describe to others what my needs were. I was just who I was, and I functioned pretty well.

It was at the very end of my PhD at the University of Cambridge when things started to wobble. It was a perfect storm: the end of a long-term relationship, a move to a new city, and a rather ambitious idea of registering to start another doctorate, this time a clinical one. As stress turned into depression, I became preoccupied with some unusual ideas. My PhD supervisor had to hold his breath as I decided to postpone the submission of my thesis. This period of my life is a bit of a blur, I think some memories were too distressing for my mind to cope with. But, rather miraculously, I bounced back. I think a major contributor to this was the support of family and friends. But I also had professional support in the form of talking therapy.

I'd like to say that's the end to this story of struggle, but the ups and downs continued. As another ending, the last month of my clinical doctorate course, loomed, my mental health deteriorated again. This time the unusual ideas became stronger, my sense of safety in the world and my trust in all others around me evaporated. After a long, drawn-out mental health assessment one day at 2am in the morning in A&E, as I grilled the assessors on their theoretical knowledge of and professional stance on whether I could be considered a 'risk to myself' (of which I kept an audio-recording as evidence in case I needed to appeal their decision), I was 'encouraged' to stay on a ward to be assessed further and supported to recover. But, again, rather miraculously, I bounced back, my unusual ideas dissipated. This time, another major contributor was medication.

It took another inpatient stay, and several relapses, for me to be diagnosed with bipolar affective disorder. I wish I could say that it has always been easy to access professional support through our national health service, and that I didn't eventually have to seek private options for healthcare. I wish I could say that everyone around me understands. Once I received the diagnosis, I was able to explain to people that under the law I have a disability;

that because of cognitive and emotional difficulties, sometimes this dis-ability gets in the way of me being able to read and write, to be organised and plan, and to make decisions quickly. At these moments I need more time, support with prioritising complex tasks, and lots and lots of compas-sion. I have now found really useful mobile apps to get round some of these challenges, and of course I have my steadfast paper diary!

Despite my longstanding diagnosis of dyslexia, and the cognitive and emotional difficulties that come with my newfound diagnosis of bipolar affective disorder, it has only been during the last year that I have started to consider that I am neurodiverse or neurodivergent (I'm not quite sure yet which is the right terminology to adopt). I think my journey into this new identity came from wearing my professional hat in neuropsychology. I began to realise that the clients I was assessing described similar experiences to me. And, most importantly, they found similar ways of coping with cognitive and emotional difficulties useful. This made me think; think really hard.

I thought, what if everyone with cognitive and emotional difficulties had access to some kind of private tutor that taught them compensatory strate-gies; what if more people made use of assistive technology; what if there was a supportive community that everyone could access; and what if everyone's employers considered reasonable adjustments?

I think that society has so much to offer people with neurodiverse needs, but we still have a long way to go. Someone asked me once if I had ever experienced trauma, and I found it hard to answer that question. On the one hand, I experienced a lot of privilege, being white and from a middle-class background, for example, and having been offered support with my diffi-culties continuously. But it got me thinking, I never really reached my potential academically early on in life and there have been times when I really haven't felt like I 'fit in'. This process can erode self-esteem. Some theorists might argue that the accumulation of 'micro-traumas' in a society that hasn't yet learned to accommodate the needs of people who are neurodiverse takes its toll.

I do not want to leave this chapter on a sad note, because I use my experiences now as a driving force to learn more and implement change in the world. I see this as my gift back.

Commentary

Sara describes how up until a few years ago she did not think of herself as neurodiverse; she knew that she had dyslexia, but even this label was not one that she readily used. She had been taught many strategies that meant her dyslexia did not interfere with her everyday life. In fact, she had suc-cessfully obtained degrees from the University of Cambridge regardless of being dyslexic.

Nonetheless, Sara is open about how her mental health suffered during periods of high stress; firstly, at the end of her PhD, then again nearing the conclusion of her doctorate in Clinical Psychology. As she details, after inpatient stays and several relapses, she was diagnosed with bipolar disorder. It is this diagnosis and the cognitive and emotional difficulties that come alongside it, that have led Sara to the realisation that she is neurodiverse. In her clinical work as a neuropsychologist, she has been able to see how many of the experiences her clients describe are similar to those she has experienced herself. She goes on to say how many of the strategies she implements in her daily life to cope with cognitive and emotional difficulties are also used by her clients.

Sara highlights how feeling like you do not 'fit in' with societal norms can erode one's self-esteem and how the 'micro-traumas' that are experienced by people who identify as neurodiverse take their toll over time. She ends by pointing out that she sees the fact that she is neurodiverse as a positive, as it helps her to give back by being a driving force to implement change.

9 Becki's lived experience

Chronic depression, long-term anxiety, and post-traumatic stress disorder (PTSD) are all considered to be forms of mental illness. While they may be invisible to some, changes in a person's way of thinking and their behaviour, to those who know them well, can be very real. These changes result from differences in how the brain, or 'neurotransmitters' in the brain, are functioning. Stripping the experience of mental illness back in this biological way emphasises recognition of this type of neurodiversity. However, noting many of the cognitive and behavioural signs of mental illness, that are akin to attentional, memory, and executive functioning difficulties seen in people with neurodevelopmental and acquired types of neurodiversity, should raise awareness too. Considering the needs of people with mental illness from a neuropsychological perspective, with the goal of living well, could significantly help to improve quality of life.

I'm Becki (yes, that's spelt right), otherwise known as Rebecca when I'm feeling fancy. I am not particularly fond of drawing attention to myself, so Rebecca is my pseudonym in a way. Use it in conversation and I won't even register you're speaking to me because, in my mind, I'm Becki. With my 30th birthday looming near, I am caught in the typical quarter-life identity crisis, but my concerns are elevated by my chronic mental ill health and the conflict I feel between wanting to be 'better' and of acceptance.

I used to be extremely hung up on diagnoses, with health professionals, often asking *why* they were so important to me. It turns out they weren't. This obsession with labels was simply a trauma response from a lifetime of having my experiences invalidated and dismissed. In the past, I've called myself high-functioning and have taken pride in being able to cosplay as a 'normal person' despite my conditions. I was effectively agreeing that my internal experiences didn't matter and being authentic to them would be an inconvenience to everyone else. To this day, I still get 'you wouldn't know you were mentally unwell!' but my reaction is different. Such phrases are often well-meant, said to reassure us that we are 'normal', but instead feeds

DOI: 10.4324/9781032714745-9

the narrative all too common with invisible illnesses; if other people can't see it, is it actually happening? If I can project a facade of mental wellness so successfully, am I *really* any different from everyone else?

With neurodiversity, I have similar attitudes to many, who for years understood it as a term purely for those with neurodevelopmental disorders like ASD and ADHD. What right do I have as someone with chronic depression and PTSD, to take up space in a community that rightly deserves more visibility and understanding? The remnants of my sociology A-level interrupted this thought, reminding me of intersectionality. Many experiences aren't directly lived by one group, and surely, significant change will only be achieved if there are as many voices as possible.

I started thinking of neurodiversity less as a construct that refers to specific conditions and instead, as a variety of the human experience. I relate to a lot of ADHD behaviours: hyperactivity noted since early childhood, difficulties in planning and problem-solving, and difficulty following social conversations. But I've realised that this is just part of the neurodiverse experience. I don't need to chase an empty diagnosis, I simply need to accept that under the medical model, I 'diverge' from the norm, and this isn't necessarily a bad thing.

My cognitive differences were easy to spot, looking back. Growing up in the late 1990s and early 2000s, mental health wasn't spoken about, especially in my family. My hyperactivity caused me to be ostracised by my peers and picked on for being 'weird', although I don't recall anyone ever explaining what it was that made me so strange. This continued until I experienced my first traumatic event in primary school. I retreated inwards, became extremely depressed, and stopped speaking in school. The hyperactivity was gone, replaced by a meek shell of my previous self. After adults overlooked the reasons for my sudden personality change, I lost another part of myself to aphantasia. I had become unable to visualise anything in my head, which evolved into a maths phobia as I suddenly was unable to do mental arithmetic. With my grades and mood slipping, I engineered a plan to move schools; away from the person who triggered my depression, hoping it would make it better. Spoiler alert: it didn't.

I continued through my school years, yo-yo-ing between episodes of severe depression, social anxiety, and panic attacks. I chose my A-levels by eliminating options – anything involving maths had to go – and flipped a coin, paralysed by the fear of failure. That's how I ended up in a psychology class, staring at a chapter on Major Depressive Disorder (MDD), and I finally felt something click. I organised a secret trip to the doctor on my 17th birthday, with my accomplices in tow, who were confused about why we were going to the doctors for my birthday meal, and for the first time in my life, I was honest about how I had felt for the last six years. I emerged with a tear-stained face, armed with the results of severe depression and a referral to CAMHS, and for once, I felt like it confirmed something I had long suspected: I was different. My friends and I went into town as per my plan, and I pretended I was fine. As I always had.

My one year in child mental health services was a mixed bag. Whilst it helped me with my social anxiety, it left more questions than answers. Questions from myself: if my anxiety is better, why am I still so depressed? Shouldn't I be fixed? And from my peers: 'Why does she always cry during presentations? She's pathetic'. This comment led to a suicide attempt that landed me in hospital. The highlight of attending an all-girls school was that conforming to the majority was integral to its social structure, and I obviously deviated from it. After I finished school, I sat with a social worker who encouraged me to go on antidepressants at 18. For once in my life, whilst I was still depressed, struggling to concentrate on anything and wanting to sleep all the time, being alive wasn't painful. I started working but quickly noticed that something in me worked differently. People didn't need to write scripts for phone calls, hyper-fixate on escape routes, or hyperventilate in crowds, and whilst I could function at work, it all came crashing down when I stepped over the threshold into my house.

It was COVID-19 that prompted some introspection, and it was then I realised that I had never really dealt with my childhood trauma and early life experiences. I developed a freeze response in early childhood, which meant when I was exposed to further trauma, I was unable to cope. I saw myself as an unequivocally broken person, someone who couldn't be 'fixed', and this attitude carried forward until my mid-twenties until I started treatment for PTSD. After multiple rounds of trauma-focused Cognitive Behavioural Therapy (CBT) and later Eye Movement Desensitisation Reprocessing (EMDR) therapy, I realised that from early life I had been conditioned to live my life finding threats in everyone and everything, which took a toll on both my behaviour and how I thought. I learnt that whilst I can actively challenge these thoughts and behaviours, they are so ingrained that I will likely be different for the rest of my life. This was when there was a palpable shift from seeing myself as disordered to just different.

I tried to chip away at my facade at work. I was known for my perfectionism, driven by a fear of failure and disapproval, underpinned by feeling unsafe. I only began to talk about my experiences after I crashed my car into a fence in a dissociative episode. I was confronted with the usual 'I never would have known' and 'You never tell us anything, so how can we help you?' which, at first, reignited the spiral of invalidation, where I convinced myself that I was simultaneously severely ill, but also faking it. Luckily, this later developed into an intense desire to advocate for myself, and a commitment to being open about how I experience things.

In my mid to late twenties, I developed a greater understanding of how my brain worked: if my body felt unsafe, my brain was too busy dealing with that rather than work, other people, and retaining information. I learnt to stop overstretching myself at work, I realised that there were certain people I could ask for help, and I learnt to do nice things for myself, regardless of my mood. Nowadays, I'd like to say I manage my conditions relatively well, but I still have to put in an extra level of effort and thought into everything that the 'normal' person doesn't.

Working in special education, there is a massive focus on neurodivergent learners and their well-being, but for some reason, it doesn't extend into the workforce. Often, my conditions aren't recognised until they impact someone else and are met with hostility rather than conversations about occupational adaptations that could be supportive. Adults are expected to manage their well-being with little consideration of how occupational adjustments could benefit them. It's only through talking with neurodivergent people that we can identify ways to improve, which benefits everyone. Having open conversations allows different voices to be heard, and the more voices, the bigger the impact. It helps us to remember that there is variety in the human experience, and the social norms we once believed in aren't always in the best interests of the majority.

I am conscious that I only represent a subset of people who consider themselves neurodivergent and, therefore, can only speak to my experiences. I have a degree of privilege in that regard, wherein I can (just about) hold down a full-time job, in a time where the importance of mental health is beginning to become more visible. But this isn't the case for everyone. Whilst there is discrimination and stigma against mental illness, society acts like these are 'easier' to accommodate than neurological and neurodevelopmental conditions. Society is tailored to the majority, and some of us can masquerade as 'normal' with adjustments to contracts, workload, and the like. But for some, even getting through the door is next to impossible without significant changes to occupational practices. The only way change will happen is if people loudly advocate for neurodivergent people to share their experiences, how present systemic practices limit them, and hear their voices. With the drive for diversity and inclusion in both societal, occupational, health and education spheres, now is a better time than any to make some noise. If the thoughts, experiences, and recommendations of neurodiverse people are not involved in policy making, then these initiatives are not inclusive.

I conclude by circling back to Intersectionality. Women continue to be disadvantaged in society; racial minorities continue to be held back by systemic inequalities; discrimination continues to affect those who deviate from the 'norm' of a white cis heterosexual man; people who identify as part of LGBTQIA+ continue to experience violence, and neurodivergent individuals are forced to participate in a world that was not designed for them to thrive. These groups are examples of people that 'deviate from the norm', but they deserve a society where their strengths, capabilities, and struggles can be considered and acted upon to drive positive change for everyone.

'Society' isn't an ethereal being that controls what is deemed acceptable and what isn't. It is the attitude of the collective. Without having space and platforms for minority groups to share and contribute, we can't call ourselves a diverse and inclusive culture. Allowing these experiences and voices to be heard, acknowledged, and validated is imperative to driving social change that improves public institutions, opinions, and attitudes. This doesn't just impact minorities, and you should be listening to these voices and, more importantly, standing behind and supporting them, regardless.

Commentary

In this chapter Becki details how as a child she experienced trauma which shaped her life going forward. At 17 years old she plucked up the courage to ask for help and although she received therapy from her local CAMHS team, she was left with more questions about herself and her place in society than answers. She goes on to describe how as she grew into her twenties, she learnt how to mask many of her difficulties, learning how to melt into the crowd so that no one would be aware of her inner turmoil. By her mid-twenties she explained that she had a better appreciation of how her brain worked; this understanding resulted in self-compassion. She is now better able to deal with her conditions, but as she points out, this takes more energy than would be the case for most people.

Becki then goes on to highlight the importance of listening to neurodi-vergent voices: without hearing from a wide spectrum of society, how can we ever say that we are truly equitable, or even aiming to get there? She emphasises the importance of everyone advocating for the needs of the neurodivergent, especially as this is an ever-expanding group of people. This dialogue and advocacy is a necessity if we are to create social change moving in a positive direction.

10 Andrew Jenkins' lived experience

Meaning-making is a way of coping with stress. It refers to how individuals construe, understand, and make sense of life events. While many view traumatic brain injury (TBI) as a physical or acquired neurological condition, it can have enormous emotional consequences. Adjusting to possible differences in appearance, abilities, losses, outlook on life, and many more changes after brain injury can be a journey, even after the accident, that evolves as the days, months, years, and decades go by. Meaning-making has been associated with a sense of post-traumatic growth, leading to new ways of thinking, living, and functioning, despite ongoing differences.

Andrew was someone whom I had known but never met before.[1] I had heard his story on a BBC TV show called *The Traitors* and had reached out to him, not expecting a response, but was absolutely delighted when he agreed to speak with me. And speak with me he did – openly and honestly, for several hours! For those of you who do not know Andrew's story, I will tell you what he told me. Andrew was involved in a car crash when he was 21 years old. He told me, 'I was pronounced dead on the side of the road', 'I went through the driver's window', 'there were five of us in the car, and I was driving'. He went on to describe the accident 'I hit a curb and I went through the window', 'I was dragged along the floor for about a couple of hundred yards until the car stopped sliding, then it landed on top of me.'

He told me how on the night of the car accident his parents got 'the dreaded knock on the door from the police to say, your son has been involved in a road traffic accident' just as they were getting ready to go to sleep. He said, 'they just assumed it was a little bump' but when they arrived at hospital 'they got taken to the side' and told 'he's not going to make it'. He relayed to me the events that followed, 'my family were told on three occasions I wouldn't make it through the night'; 'they said if he wakes up from his coma, he's not going to be able to walk'. But Andrew, a man who learnt about bravery and fighting spirit through being a talented rugby player in his youth, was not to be discouraged! He told me, as a patient in

DOI: 10.4324/9781032714745-10

hospital after the accident, his mother had told him: 'it was embarrassing; you were lying in bed with all the tubes in you and you are arguing with the surgeon, the doctor, telling him you are going to walk!'. Despite everyone's predictions about his prognosis, Andrew proudly said 'It took me two weeks to walk again'. He reminisced: 'the first time I stood up with the physiotherapist, I couldn't use my muscle memory,' 'I couldn't find my leg. I stood there. I just couldn't remember. I could not move my legs. I knew what to do, but I could not move my legs.' All I could express at this point in the interview, after seeing Andrew bound into the meeting with me, was awe. I remembered that Andrew had recently set up a company called 'Strength in You' and I truly understood what this meant at that moment.

When I asked Andrew about what he thought of 'neurodiversity' in the context of his experiences, he laughed and said, 'it's obviously a buzz word at the moment!', and then more sincerely 'I probably fit in there because after my car accident, I suffered a severe brain injury'. He told me that he'd mostly heard of neurodiversity used in the context of ADHD, and reflected that he could identify with ADHD himself, even though he wasn't quite sure he could explain why he felt this way. But most importantly, what I could see for Andrew was the fact that he saw this as a very positive thing: 'I talk very fast now; and I get so many ideas and creative thoughts'. He said 'It's a good thing to me. It's not a bad thing. It's quite a big talent. A lot of famous people, successful businessmen, have got ADHD'. He went on to say, 'I know there's different ways people learn'; 'they might not be great academically, sitting looking at the board in school, but they're so creative; they've got so many other skills that companies can utilise.'

My discussion with Andrew on the topic of neurodiversity sparked a further conversation on mental health. He disclosed that after the accident 'I struggled for 20 odd years with my mental health', 'I'd get up every day for a corporate job, put the shirt and tie on, a big smile on my face but inside I was broken and sometimes I'd lie in bed crying'; 'wished I wasn't here anymore, hated the way I looked in the mirror, call myself a freak every single day, it affected my relationships. I jumped from one girl to another, to another. I hated the way I hated myself. I hated the freak. I hated to look in the mirror'. He went on to say, 'I was trapped between the sixth and seventh stages of grief...for years', 'I was really depressed, bargaining, all this sort of thing'. He reflected on how, at times, even before the accident, his emotional experiences seemed out of control. He told me 'I read a book called *The Chimp Paradox*'. This is a book that details a psychological model by Dr Steve Peters that explains the inner conflict within our minds using a simple analogy involving three main components: the 'human' which represents our rational, logical thinking, the 'chimp' which symbolises the emotional, impulsive part of our brain, and the 'computer' which stores memories and automatic behaviours learned from past experiences. Andrew said to me, 'I think my "chimp" was running my life for 40 odd years. It was running my life; it acted on feelings and emotions. But my logic, my human brain, was quite weak, I suppose'.

It has been a journey for Andrew to reach a different stage in his life. He explained how looking after himself physically (exercising and eating well) and mentally (through meditation and brief psychological therapy) has helped him to adjust and control his emotions and actions better. Having people around him for support has contributed, and how leaning into vulnerability has been both scary and liberating. He explained 'We think vulnerability is weakness' but actually it's 'really brave and courageous'. He went on to tell me 'I did a talk yesterday', asking the question 'who's been vulnerable?' to the audience; everyone put up their hand, until he asked, 'Who liked the feeling?' and everyone put their hand down. Andrew described how he read out the following statements about real-life scenarios: 'a man helped his 42-year-old wife in her last will and testament who's been diagnosed with stage 4 breast cancer terminal; a man knocking his best friend's door telling his only son has been killed in a road crash; somebody being pregnant after multiple mis-carriages; a person saying I love you in a relationship and hoping somebody says it back'. He then asked the audience 'do you think they're all weak?'; and guess what? No one did.

For Andrew, it seemed apparent to me during the interview that his life experiences have made him a stronger person. He told me, 'I met the man last year who saved my life, and he called me a miracle patient', the doctor said: 'You should be dead'. 'My moods are a little more volatile now', he acknowledged, but he also spoke about the concept of 'post traumatic growth'. He likened his experience to others who have lost limbs in accidents but go on to become successful paralympians. Andrew explained 'they've turned it around, they're using the trauma for good', 'they're doing great things off it'. Since my interview with Andrew, I have seen him appear on social media many times telling his story and motivating others to gain something strengthening from their life experiences too. His view is that education is one of the most important things to raise awareness of the needs of people with acquired neurodiverse conditions, resulting from brain injuries. He told me, 'People who have suffered a brain injury are different in some ways, but they still may work exactly the same in other ways. They might have to work in different environments but people, bosses, CEOs, MDs, need to understand it better, not be frightened by it, we need to lean into it, embrace it. Like, you know, employ people who are neurodiverse', 'employ these people and understand it better'. Andrew felt very strongly that 'I don't use the word disability', 'It's not disability, because disability is when you can't do something' and 'I'm an able-bodied person who can bring lots of benefits to the business now'. Alongside this agenda, he has an interest in gender equality, and the mental health needs of men as well as women. My parting thoughts are that Andrew is a force to be reckoned with, and he is determined to change society for the better – watch this space!

Commentary

Andrew's story is one of determination and will. He is an individual who, although having experienced a life altering event, continues to work towards his ambitions. In his own words, he does not see himself as disabled, rather as someone who has an awful lot to give to society. He is able to acknowledge the tremendous effect the injury caused, both to his physical body but more importantly to his mental well-being. Andrew describes how at times he would despise his new body and would wish that he had not survived the accident. Nonetheless, despite having these dark moods, he has also been able to turn his life around, so much so that he was a finalist in the BBC's production of *The Traitors*.

Turning to his thoughts on neurodiversity, it is interesting to read that the title 'neurodivergent' does not leap out at him when he describes himself. Instead, he acknowledges that having an acquired brain injury will result in some divergence, but that he sees this as a positive. He highlights how many successful people are neurodivergent, especially with ADHD, and goes on to discuss how it is important that businesses take this into consideration. Andrew points out that people need to be educated around brain injury and neurodivergence so that they are less scared by it and understand it better.

Note

1 Andrew was interviewed by Sara Simblett.

11 D's lived experience

When do we ever really know what is going on in someone's internal world? We often rely on self-report or performance-based measures that give us some insight into one specific moment in time, but thoughts and behaviours are constantly changing, and that is part of the beauty of diversity in human experience. When given the diagnosis of a neurodevelopmental condition such as ADHD from a young age, this can have a profound impact on the way a person grows up to perceive themselves, and the emotional impacts can be far-ranging, from acceptance to confusion to rejection of parts of oneself. While knowledge of differences can support an individual to thrive if they are put in the right environment, labels and diagnoses can change people's sense of who they are and there becomes a responsibility in society to respond and adapt to this.

My name is D. I'm a PhD student studying psychology; as of writing this, I am 25 years old, I like to paint and make art, play guitar and write music, read, and hang out with my cats. I have been diagnosed with ADHD.

I've chosen to share my experiences because I believe that, especially within the field of psychology, too much emphasis has been placed on understanding neurodivergence from a neurotypical perspective, with neurodivergent individuals being spoken for and spoken about in ways that have historically stripped our population of autonomy, and often dignity. It is only a very recent development that within psychology a neurodivergent researcher can stand on equal footing with a neurotypical peer, and have their opinion, let alone their work, viewed as credible. I would like to use my voice and position to help highlight the importance of our own perspectives shaping the research paradigm, rather than addled assumptions made about us, and for us.

I find it difficult to focus on specific anecdotes that highlight times that I have realised my ADHD was impacting my life or having revelations about myself in relation to my neurodiversity. I find myself unable to separate my neurodiversity into specific experiences or moments. I was diagnosed as a

DOI: 10.4324/9781032714745-11

child, so have always viewed myself in the context of being different from my peers, although even before my diagnosis I could feel a disconnect between myself and that of most kids around me, and a kinship with the kids that also seemed 'othered'. I would have an intuitive feeling of 'Oh, you're like me'. I guess that could be seen as an experience. But throughout my life, I have never perceived my ADHD as an illness, it has always simply been part of who I am, how I think, how I conduct myself, and how I live my life; it is inseparable from myself. I think this is why I have always been quite critical of a notion of 'curing' a neurodivergent condition. However, there is something to be said about neurodivergence and disability, and while I do not view my ADHD as an illness, it can certainly be disabling. Society is not tailored for us and is often quite hostile to the needs of those with neurodivergence. I have been on ADHD medication for most of my life, and I do not believe that it has 'cured' me. Rather it has assisted me in participating in the requirements of society, namely productivity. Yes, I am more able to conform to the social and work expectations of the neurotypical world, but this is a façade. Ask any neurodivergent person and you will likely hear a similar account and hear the term 'masking': to wear the face and act the way that you believe is expected of you. Neurodivergent people do a great deal of work to understand and relate to neurotypicals and attempt to act 'normal' so as to not make them feel uncomfortable. I only began to question in my adult life why the same is rarely afforded inversely. As public awareness of neurodivergence has grown, so has neurotypical understanding and patience for our needs, so I am glad to see this gradual change take place. I've gone on a tangent: I suppose the experience I want to centre upon is this feeling of misunderstanding that has remained a constant in my life.

I believe I first began seeing the term 'neurodivergence' used on the internet, often by autistic and ADHD individuals creating long threads providing information about their conditions, often in educational or autobiographical formats. I remember seeing the term for the first time in maybe 2014 and feeling a sort of unity and communal understanding of myself in relation to a group. It made me feel less alone in my experience, and more aware of our ability to self-advocate and organise as a community, where previously I had felt helplessly at the whims of those with authoritative control over my needs and the accommodations I was afforded for them. These threads also provided the language and terminology needed to identify the aspects of my own experience, and help me to verbalise my issues, and what I needed to accommodate them.

I feel quite strongly about opening up the term 'neurodiversity' to be inclusive of a range of cognitive differences, not specifically those with ADHD and/or autism. I think the nature of neurodivergence as a term is inclusive to anyone whose mind works differently from the 'average' person. I strongly advocate for more importance to be placed upon individual differences, not just rote categorisation. ADHD is just a label to help group a series of traits: no two people with ADHD will exhibit the same traits in the same ways. If we define neurodivergence from a basis of understanding that there are people whose minds work differently from those of others, why wouldn't we allow self-identification with that label to anyone who feels it is applicable to them?

Personally, I have always had a bit of an odd learning style that has not been overly conducive to structured schooling. I find I learn best when I am provided with questions, and the resources to research those questions independently. I read faster and with more ease than I can listen. I think I have always struggled with this in school settings, as having information slowly explained and drip fed over weeks and months would mean that I would struggle to absorb it while it is being spoken to me, and by the next class I would have forgotten what we had covered in the last, and how it related to the previously covered material. Growing up, I had quite severe migraines, multiple per week, which would debilitate me for the entire day, meaning I missed quite a lot of classes. From when I was quite young, I had to spend much of my time catching up on material I had missed and effec-tively teaching it to myself. By the time I reached high school (I am Cana-dian), I found that I learned best by receiving an overview of the class and its learning outcomes and reading through the entire textbook and taking notes as I went. I would do this before classes would even begin, allowing me to ask questions and refine my understanding through the classes throughout the year. However, it often meant I felt under-stimulated and unchallenged by my time in class. It was a common theme; my teachers would voice worries about my lack of attention and restlessness. They would find me drawing instead of taking notes, only to later be surprised at how well I would perform on graded material. Throughout my adult life though, I became more aware of challenges that I hadn't previously noticed, mostly with executive function and cognitive flexibility. I think the greatest hurdle I've had in adulthood has been my ability to organise and manage my time effectively. I find it challenging to switch between tasks: whatever my mind is attending to at a given moment is what I can do, and because of this I have struggled with prioritising parallel tasks. I tend to have a recency bias when making these internal judgements, which can be quite frustrating when I have previously convinced myself of an order in which to do my tasks. If a new task is dropped on me after that, I can get overwhelmed and feel unable to do any of them. Then later, when I feel able to work again, I'll have forgotten the previous internal prioritisation, only remembering the most recent task. This means I often forget the things that I need to be doing, which in many cases are quite important things that in forgetting triggers new tasks to compensate for having forgotten. This can create quite an ugly cycle, where I end up feeling like I'm chasing my own tail a bit, creating more work for myself to make up for the work that I had forgotten, which causes me to forget other work, and so on. All this considered, sometimes I am just flat-out unable to do the things I need to do, no matter how badly I need to. Sometimes I feel completely incapable, even if it is a task as simple as getting up to go drink some water; I can end up feeling stuck, knowing that I should do it, and wanting to, but unable to have my mind convince my brain to convince my body to actually do it. This of course applies to my work and productivity, but also my social life, my personal health, and wellbeing, everything really.

My cognitive differences definitely change over time, sometimes daily, sometimes in longer periods of weeks or months. It is challenging to pin-point, but I think they change reflectively to where I am more generally in my life. If I am unburdened and stress free, having my social and emotional needs met, my financial and material needs secure, I feel more capable; but that could be said of anyone. I think my cognitive differences can worsen in times of difficulty, spurred by external factors outside my control. It can often feel like my own brain punishing me for struggling, which in turn causes me to struggle more. This can manifest in self-resentment, and an internal narrative where external factors exacerbating my issues are some-how my own fault. Most of how I plan my life now is directed at making sure I have built safety nets and supports to fall back on if unexpected stress or duress occurs, so that if I am overwhelmed, I am able to quickly return to a baseline, rather than spiralling into burnout.

I ebb and flow through familiar patterns. Generally, my pattern is a con-stant tug-of-war between needing novelty and needing routine. For example, I find that I can easily understand plans in a macro scale: I can look for-wards and plan into the future. But I can struggle with the micro scale: the planning from day-to-day to achieve those larger goals. I often feel like I'm dealing with a different set of mental tools on any given day. I've coped with this by trying to teach myself to be habitual, and to enforce structure on myself. I have a condition called circadian sleep phase disorder, meaning that my body is not on a typical 24-hour sleep and arousal schedule. Mine is 36 hours which causes my sleep schedule to gradually change throughout the year. I think I struggled more as a child and teenager with this than as an adult, as I am now lucky enough to be in a work environment where I usually determine when I start and finish my workday. When I was younger, it was gruelling to be perpetually forced into 7am classes when I often hadn't even slept. But this lack of routine is pervasive in my life. I've tried finding weekly activities like sports and regular social engagements to start building new habits. I think this lack of routine has resonated in many of my hobbies as well and is why I don't always stick with a hobby for long. I may learn a lot about how to do something initially – play an instrument, for example – and become infatuated with it for a period. As time goes on, and the infatuation fades, I lose the habituality of practice and become frustrated when I hit a wall with my progress that can only be overcome with regular practice. Nowadays, I am trying to take things more slowly, allow myself a set amount of time per day to practice something that I'm trying to improve in, and try to stick with that. I haven't necessarily succeeded with this method yet, but it's where I'm at. However, I crave novelty. I even struggle to re-watch a movie I've previously seen, as I already know what happens. As an aside, I believe I have quite a strong long-term memory, and a somewhat weak short-term memory. Of course, I enjoy familiarity, but I perpetually desire something new. Think about going to your favourite restaurant and getting your favourite dish, you know you like it, but if you have it every time, predictably on the same day, same time each week, that enjoyment would become predictable would it not? I enjoy knowing that my favourite things are always there to return to, but I also need to try new things. Whether this is a certain subject I am interested in learning more about, or a certain activity, or a friend, anything really. This is also reflected in my everyday habits. If my day is predictable, I feel unengaged and apathetic, and if I feel this way, I am not going to function optimally. Like I said, there's an ebb and flow, before swinging too far to one side and rebounding to the other.

Others have told me that I can come off as quite aloof, transient, sometimes I've even been told I come off as mysterious. In reality, I think I am often just internally preoccupied. I can forget social engagements are happening or lack the executive function to actually begin the process of getting up and getting ready to go to something even if I want to go – this happens quite often and usually causes me considerable distress. Otherwise, I am regularly told that I am perceived as not paying attention or listening to others, which is sometimes true. Somebody might say something to me that triggers a thought that I attend to, and in the process, I will have missed the rest of what they were saying. Other times though, I will be listening, but perhaps because I will listen silently, or not look directly at them when they speak, they will think I didn't hear them when I actually did. I think part of this comes from their familiarity with me and knowing that I will sometimes 'tune-out', and it can be difficult for them to tell whether I've stopped listening or not. I've also been told that I struggle with switching tasks, which I think I touched on earlier. If someone tells me to do a new task while I am already doing something, it can overwhelm me a bit and cause me to feel incapable of doing either. Sometimes I will try to finish what I am doing in the moment before switching to the new task, but this can mean that I am perceived as not acknowledging the urgency of the new task, or hours might go by before I finish my initial task, and in that time the person will think that I have forgotten the new task and become frustrated with me, even though I hadn't forgotten it. Though sometimes I do just forget.

To cope with my cognitive differences, having specific systems in place to help me. I struggle with my memory at times, so if I create a sort of shorthand with myself, I can make things a bit easier. For example, as with many individuals with ADHD, I commonly misplace items. I will always know where I put something if I have assigned that item a location, my wallet and keys always go to this place on my desk, glasses go in this spot, and so on. This is one example, but I think it extends to many other things as well.

Basically, creating systematic rules for myself to get through tasks that others might do intuitively. Issues can arise though when other people don't know my unspoken rules for how specific things should be done.

It can be difficult for others to help I think, as no neurodivergent person has a one-size-fits-all way that they would want help or support. Generally, the easiest way is to ask. If I find it difficult to pay attention without having something to do with my hands, ask if I need something to fidget with. If it seems like I didn't process what you said, ask if I need you to repeat it. Things like that I suppose. I think ideally, a neurodivergent person should tell you about their needs for a given situation, or what you could do to help. For me specifically though, I'm not sure. I think I struggle most with executive function and actually starting tasks, so it can be useful when somebody starts a task with me so I can continue and finish it on my own.

Again, if we're looking at how society more generally can accommodate and support neurodivergent people, I think being able to ask questions, and for neurodivergent people to make accessibility requests without fearing backlash for potentially inconveniencing their peers is critical. Neurodivergent people do an awful lot of shaping to the needs of their neurotypical peers, and learning how to adapt to their needs, having the same afforded to them would go a long way. I believe that needs to come from widespread education and removal of stigma, even though I have seen the general public become more aware of symptoms, terminology, and the existence of neurodivergent people, I have just as commonly seen the same people preaching for acceptance for being annoyed at neurodivergent people that inconvenience them or gross them out for not behaving as benignly as they had expected.

Honestly, it's difficult to pinpoint what can worsen the detrimental aspects of my cognitive differences: I'm not totally sure. I would say stress would be the most acute indicator I have for when my cognitive differences become more pronounced. I often think of my mental state like a cup full of water, and stress pours more water in that cup in different amounts depending on the severity of the stressful situation. If my glass is already about to overflow, even a single drop of water can have the surface tension break, and it pours over. I hope this metaphor makes sense. I think neurodivergent people are typically quite noticeably perturbed when something has stressed them out or overwhelmed them, and often are told to just push forward, get over it, stop being so sensitive, etc. I think that is what others do wrong most often. If a neurodivergent person is overwhelmed, allow them the space to decompress, because otherwise I think that state of overwhelm can just snowball.

Being able to compare your experience to others is a very human behaviour I think, and is especially important for neurodivergent individuals. I have a lot of neurodivergent friends, all of who have different cognitive differences of varying intensities, so it is easy to see similarities and differences. I think there is generally a level of commonality that neurodiverse people share, which is often why we group together and so easily befriend each other, but at the same time, you can sometimes see 'bad pairings'.

Colloquially I've heard 'incompatible types of autism' said for when two people are just too different to interact comfortably, as they both overwhelm the other with their specific quirks. I think I relate more to those with ADHD than autism, though I can see how I have a foot on each side. I think neurodivergence is not a clear-cut thing where once you have your diagnosis that fully accounts for who and what you are. I also think that I have had a greatly different ADHD experience from many others, because I was put on medication so early in my life, while many others I know were not. I had therapy and pharmaceutical interventions that helped me to develop many of the skills needed to cope with my cognitive differences, and while I can relate to the struggles of my ADHD peers, there are certain cognitive strategies I have developed that were ingrained into me so early that I can often forget that others do not have the same toolkit at their disposal that I do.

Generally, if I was thinking of how the world could be more accommodating to me, I think that having society on a nine to five workweek is a pointless endeavour that benefits nobody. There are so many other systems that have been shown to increase productivity, efficiency, and wellbeing; it is antiquated, and I think would benefit everybody, but particularly people with ADHD, myself especially.

Just as a final takeaway thought for the reader. Neurodivergent people are just like anybody else, they just think differently than you. But you think differently from everyone else as well. I think the world would improve a great deal if we approached each other knowing that we all have individual differences to accommodate and stop trying to create a one-size-fits-all world. Like any good relationship, the best thing that you can do is communicate, ask questions, find out what makes others comfortable and uncomfortable, and try your best to accommodate to their needs. Finally, be willing to mess up and don't be hard on yourself if you do. I promise you that a neurodivergent person will be happy that you are trying to accommodate them rather than upset that you didn't accommodate them correctly.

Commentary

In this chapter D explains how their diagnosis of ADHD has affected them throughout their lives. D eloquently describes how having ADHD has shaped them and their experience of the education system. They go on to describe how they have learnt to implement strategies that help with the everyday cognitive challenges that they have experienced because of ADHD. They highlight the importance of early diagnosis and treatment, whether this is through pharmaceutical intervention or cognitive therapy. D appreciates how many of the strategies they use to mitigate much of their cognitive difficulties they have used since childhood so often these are ingrained: this is not the case for all people with ADHD.

D describes how they do not see ADHD as an illness to be cured, but rather a part of who they are. D continues to explain how society can make the diagnosis of ADHD disabling. This occurs either through a lack of understanding or willingness to make relevant adaptations. At times they describe how this can make a person with ADHD feel an inconvenience. D points out that many neurodivergent individuals find themselves either masking their differences or shaping themselves to the neurotypical world. They are rightly critical of both of these changes that are made by the neurodivergent individual to 'fit in' with society.

D highlights how the term neurodivergent should not only apply to those with ADHD and autism but rather should apply to anyone who's mind works differently from the average person; this is in keeping with the current theme of this book.

12 Patrick's lived experience

Multiple sclerosis is, in many cases, a neurodegenerative condition, meaning that abilities associated with the brain and central nervous system functioning progressively decline over time. Some of these changes are more visible than others, with many people finding it hard to talk about the loss of more personal and invisible functions. Holding on to one's dignity throughout these multiple losses is important, but takes strength, courage, and a lot of energy. Maintaining a sense of independence is often a sought-after goal. Individual neurorehabilitation, such as occupational therapy, physiotherapy, and psychological therapy, can help to achieve this but living well with MS often also requires the collective care, understanding, and compassion of others.

I have known Patrick for some time now.[1] He has been instrumental in guiding research conducted on the topic of multiple sclerosis (MS) for many years. He is an advocate for the needs of people with MS and regularly participates in the teaching and recruitment of staff in clinical services within the UK National Health Service (NHS). He agreed to speak with me about his personal experiences of living with MS, as well as his professional take on the subject of 'neurodiversity'. We met on a video call, and he started by saying 'we are hopefully making society as accessible as possible, but people are going to have different problems'. In relation to himself, he told me 'My thinking has definitely deteriorated'. He went on to explain, 'I can't do two things at once', telling me about a recent meeting he attended where he was in a room with a lot of echoes, and how he found it incredibly hard to concentrate. He is very frank and pragmatic when he says, 'I would not have had any trouble in the past... now, I'm easily distracted', and caveats his own personal experiences with 'but each person with MS is affected differently, because it affects different parts of the brain'. He explained that for him, not only is cognition affected but also movement 'because the message is not getting down my nerves to control the fine control of writing and I've got bladder problems, bowel problems, and mobility issues, I've also got double vision issues, and I get stressed very easily'. Patrick, for as long as I have known him,

DOI: 10.4324/9781032714745-12

has always been extremely practical and motivated to overcome challenges. He told me 'I like things to be organised; if something out of the ordinary happens, I get really stressed about it.'

Patrick has not always had these needs, as he explained to me that his symptoms of MS have 'gradually become more and more significant and affected my life to a greater and greater extent as the MS shreds nerves'. Patrick is clear at this point that MS can affect any part of the central nervous system, including nerves in the spinal cord and brain. Patrick feels strongly (and does not shy away from any controversy) that MS is a progressive condition on a spectrum. Although MS affects people in different ways, rather than progressive stages being something separate to relapsing remitting stages, Patrick's view is that it ultimately leads to a deterioration in everyday functions.

When I asked how Patrick copes with this, he answered 'it's a case of learning to understand what I can achieve and what I cannot achieve and not trying to overachieve because then I will not do anything.' He describes how life has had to become a lot slower paced due to MS. He tells me how the hardest part is the social isolation that he experiences: 'your social capital shrinks', he said. He comments that living in urban busy areas can be worse than in smaller rural towns. In the capital city where he was living at the time of this interview, Patrick said with sadness 'it feels as if I'm looked straight through, people don't see me... I'm just not there, I mean, they'll be all around me, but they don't notice me; they don't try and engage in a conversation; if they see me struggling to get something off the top shelf in a shop, they continue what they are doing; it's the exception who will say, can I help you?' He went on to say, 'I feel here where I live, it's a rush and people want to get from A to B as quickly as possible, but when I lived in a rural area it wasn't the same, people stop and have a chat.' We agreed that it sounded like what Patrick was describing was a lack of community in some areas, and that lack of community makes his life different, probably harder. Many a time my eyes have been opened when Patrick recounts the difficulties he experiences due to his physical needs, including the fact that he uses a wheelchair – the cobbled streets, the steps into shops, the unkempt pavements with dangerous bumps and holes, the gap between platforms and trains, the barstools and high tables out and about.

But, ever the optimist, Patrick explained 'I try to keep myself alert.' He works hard to overcome the challenges he faces. He tells me that journeys, for example, take a lot of planning: 'I might spend an hour working out how to get from A to B, then contact the right people to help me.' He smiles at this point when he tells me all about the 'perks' open to him in society, such as the discounted theatre tickets. If only people knew the sometimes frankly frightening experiences that he has to undergo to travel to do something enjoyable. He sighed when he said, 'it takes quite a lot of time and energy and processing I suppose'. This doesn't stop Patrick, as he explained 'being constantly busy is just a family trait of mine, my brother and sister are the same!' And he reflected on the fact that 'not everyone has got that drive'.

Patrick is very invested in making a difference in the world, but he told me, 'It's no good going in moaning to the nurse in the hospital. You've got to go to the people at the very top and they can then cascade it down. Because it's coming from a senior person, it's got a lot of weight attached to it.' He explained, 'the same goes if you have problems on a bus. You don't take it out on the driver. That's just going to make the driver really cross and upset everyone. But you can go find out the directors or whoever and you can complain to them about the problem.'

On the direct topic of neurodiversity, Patrick had strong opinions too! He told me 'I'm convinced there are different types or classifications of neurodiversity, in the same way that not everyone has the same cancer.' Prior to this point, Patrick had mused 'I think this discussion is intensely frank and open, we haven't beaten around the bush!' And the relative controversy of his comparison to a diagnosis of cancer didn't go unnoticed. He explained to me 'there's a bit of context that's needed in addition to the term neuro divergence or diversity; it needs to be said whether it's acquired or whether it's from childhood and things like that.' He also pertinently put, 'the medical landscape for all conditions changes', and he went on to describe how scientific understanding and treatments evolve over time and that this needs to be considered in the context of using the term 'neurodiversity'. Wise words that I will certainly reflect on.

Commentary

Patrick displays tremendous strength in the face of a deteriorating health condition. Unlike many other people in this book, his future is more uncertain as the MS continues to degenerate his brain and nerve fibres. In his interview he is candid about the fact that his life has changed significantly, giving examples of how his cognitive abilities have lessened over the years, as well as discussing how his mobility is now an everyday challenge. Patrick describes how having this condition can be very isolating, especially as his disability becomes more pronounced.

Nonetheless, Patrick continues to strive to live his best life, which includes many things that people without a disability do not do. He has learnt strategies to overcome many of his difficulties and has learnt the importance of taking time and planning for events.

On the topic of neurodiversity, he feels that this is still in its infancy within the medical field. He points out that there should be different classifications for neurodiversity depending on whether it is acquired or from childhood. He finishes by pointing out that scientific understanding evolves over time and so should the framework for using the term 'neurodiverse'.

Note

1 Patrick was interviewed by Sara Simblett.

13 Joe Kelly's lived experience

According to the Diagnostic and Statistical Manual of Mental Disorders (DSM), ADHD first appeared as a diagnosis in the second edition (DSM-II) published in 1968, where it was called 'hyperkinetic reaction of childhood'. It wasn't until the 1980s that it began to be considered to be known as 'attention deficit disorder, with or without hyperactivity' and later 'attention deficit hyperactivity disorder'. This does not mean that ADHD did not exist before this point, but our societal understanding has only really been forming for the later part of the last century. Overlap with other diagnoses can be common. One such overlap is with 'restless legs syndrome' (RLS). There are thought to be shared underlying mechanisms related to dopamine levels in the brain. Because of these overlaps in the way society has conceptualised the difficulties people who identify as ADHD experience, misunderstandings can occur, so it is important to keep an open mind in the context of an ever-evolving understanding of why people are the way they are.

My name is Joe, and as I write this, I am 39 years old, recently unemployed, and soon to attend the graduation ceremony for the creative writing Master's degree I just completed. The concept of 'neurodiversity' only recently came into my life when I was diagnosed with ADHD at the age of 36. My only prior image of the condition was the classic movie caricature of a loud, disruptive little boy, which seemed very far from the introverted person that I have always been. The more I learned about it, however, the more apparent it became to me how little any of us really understand about ourselves and how we function. I have always believed that one of our greatest and most fundamental strengths as a species is our ability to communicate and share our experiences with one other. It is what helps us connect with the people around us and form communities that provide invaluable security and support. Within these communities, pictures of the past are painted and its stories told and retold so that the lessons learned by each individual can be learned by all. It almost allows us to experience things not as a single entity, but a collective. As we carry our own past

DOI: 10.4324/9781032714745-13

experiences into the next one, we also carry the experiences of every other person we've interacted with or learned from. It's what makes it so we're never really going it alone. It is truly an amazing thing, but it has been falling short when it comes to the problem of expressing the experiences of people who fall under the heading of 'neurodiversity'. I have found it extremely difficult to write this chapter because ADHD has affected me in so many ways that there isn't room to explain it all here. What is clear to me is that ADHD is not just something I *have*, it is a fundamental part of my being. It isn't the storm that assaults the fortress of my mind, it is the stones and wood that the fortress is made of. Instead of attempting to dissect all these effects here, I'm going to share a smaller example of something from my life that I think illustrates a lot of the themes I think are important.

When I was a child, getting me to sit still was a terrible ordeal for everyone involved. My mother is a hairdresser and still mentions sometimes how much of a nightmare it was to get me to sit for a haircut. I would squirm and cry and she would try everything from threats to bribes to stop me, but nothing worked. I remember being asked why I couldn't *just sit still*, and all I could come up with was to say, 'my feet feel funny'. I can only imagine how exasperating it must have been for my parents, but I simply couldn't explain myself any better. They even took me to a doctor who asked all the questions I had already been asked before. Was it itchiness? Was it pain? Was it stiffness or weakness? The answer was always no. As far as I'm aware, no conclusion was ever reached. Eventually, I did get better at getting through haircuts without complaining, but it wasn't because the problem went away – it was because I stopped talking about it. I can't even articulate it properly now because if a word exists for the unbearable sensation that still keeps me awake at night, screaming at me to *move*, it isn't in my vocabulary.

I have since discovered the existence of 'restless leg syndrome' and found some measure of validation, but it still does nothing to solve what I think is the main problem in that story, which is that there is a disconnect between people that happens when they don't have shared experiences. When we use words to communicate something, what's really happening is that a connection is forming between our experiences that enables us to understand each other. An example of a strong connection could be me expressing to you that my leg is 'itchy'. What makes this a strong connection is that when I write the word, I am expressing a specific sensation that is common to both of us, so when you read it, you're able to connect the word to the exact same sensation recalled from your own experience, which is what brings understanding. This connection process breaks down, however, when either side does not have the required experience. The problem I was having in the restless legs example, and have had in many other situations in my life, was that I was trying to communicate an experience that was not shared by the people I was trying to make understand me. As far as everyone was aware, the experience of life was the same for everyone. If what I was trying to explain wasn't something that existed in their own experience, they would struggle to form an understanding because they would have nothing solid to connect it to in themselves. It's like trying to paint a kaleidoscopic, constantly morphing shape with a paintbrush far too thick for any detail, for an audience that only sees in black and white. I think this is why people with ADHD so often favour the use of analogy and metaphor when trying to explain things. We're so used to our experiences not lining up with the people around us that it becomes habitual to come up with creative work-arounds to express ourselves in ways that others might be able to relate to.

As far as the terms 'neurodiverse' and 'neurotypical' themselves go, my feelings are still forming, but I do have concerns. I can see that in some way it is an effort to recognise that there are smaller groups of people who experience things differently from the majority, which is an important and valuable pursuit, but in this case, I don't think it works very well. When the two words are used in tandem as opposing categories, then I don't see how it's any different from just saying 'abnormal' and 'normal'. It also doesn't really make grammatical sense because it's like saying 'you're either a normal person or a diverse person'. It's *people* that are diverse and what we call 'normal' is just a subcategory within that. The way I've come to think of people is something like how I imagine an experienced carpenter thinks of wood. If you went to one and said you were building something and wanted to know a good wood to use, before anything else, they would have to ask the question: 'what are you building?' They would understand that, just as trees adapt to different climes, wood itself is highly diverse. When approached with a project, they don't just throw any old wood at it, they use their knowledge of it to thoughtfully select the one that has the properties best suited to the job. I think that if we understood ourselves and each other in this way, then things could run a lot smoother. At the moment it feels like our understanding of ourselves as a species is more like someone who keeps thoughtlessly cobbling things together out of random scrap and wondering why they always fall apart. We lack the understanding because we've barely even taken the time to look properly at what's in front of us. If we took time to do this, we'd see that we have far more at our disposal than we realise. Then, once we've taken a proper inventory, we can undertake the real work operating at a much higher capacity.

It definitely seems like a monumental task, but I'm not without hope. I know there's nothing I can do for my own younger self now, but one thing I would really like to see is more effort being made in schools to help each child at least understand *themselves* better. That way they would be able to learn their own strengths and weaknesses and make a more informed choice on how to apply them to their lives. For me, the biggest positive I have felt with all this is the validation that comes from finally knowing that all my feelings of disconnection throughout life were true, and that I'm not actually alone in it. It sometimes feels like trying to explain myself is like banging my head against a wall, but when I finally had the experience of connecting with others who could really *understand*, it did more for me than I could ever articulate. This is why I wanted to contribute to this book. My hope isn't to reach everyone or bridge an impossible gap. Instead, it's to keep sharing my experience so that one day it might provide that feeling of connection to someone else. I want people to realise that it isn't actually a wall they're banging their heads against, it's a door, and more of us are finding the keys to open for each other.

Commentary

In Joe's chapter he highlights the importance of finding a common language to describe the neurodivergent experience. He acknowledges how not everyone who identifies as neurodivergent will have the same language but also highlights that all these languages will deviate from the neurotypical. He describes how having a form of communication that leads us to understand how the other feels is important, but that this has been lacking for those who are neurodivergent. He explains how trying to communicate an experience that is not shared by most people can lead to struggles in understanding each other.

Joe has issues with the terms neurodiverse and neurotypical, pointing out that it is just another way of saying 'normal' and 'abnormal'. He also highlights that diversity is part of being human, so in this sense everyone is neurodiverse. He ends his chapter by saying how more education at school level is necessary to help each child understand themselves, their strengths, and their weaknesses. With this, hopefully a sense of validation will be achieved earlier on, as the individual realises that their feelings are true and that they are not alone.

14 Lisa Beaumont's lived experience

A stroke is a type of acquired brain injury (ABI) that is caused by either a blockage or rupture of a blood vessel within the brain, which causes damage to the surrounding brain tissue due to a lack of oxygen available and, in the case of ruptures, pressure within the skull. Strokes can be a life-threatening neurological condition. With treatment, however, the damage to the brain can be contained. The long-term effects of a stroke, after treatment, somewhat depend on where the stroke occurred in the brain, but can lead to 'hemiparesis' or weakness or the inability to move on one side of the body, as well as cognitive and behavioural changes, including severe fatigue. Living with these differences can be difficult, especially explaining to others the impact of such consequences, but that is not to say that people cannot live a full life after experiencing a stroke either. Finding ways to maintain a positive outlook and live in line with one's values is a common goal set by many in neuropsychological rehabilitation.

One of my first questions to Lisa was, 'do you want me to call you Lisa in this book?'.[1] Lisa laughed and replied: 'I'm very happy for my name to be mentioned!'. She went on to explain that in the past, 'I would have been much more cautious and private prior to the brain injury, whereas now I'm very happy to share my story'. Thirteen years ago, Lisa experienced a cerebral haemorrhage (more commonly called a 'stroke'). She describes herself as an 'active person' and was both a successful businesswoman, working in marketing, and a mother of children who were in primary school at the time.

Lisa explained that she now lives with multiple disabilities. She told me, 'I'm left side paralysed, so I can't walk or do cooking or anything that uses two hands, and I'm also partially sighted due to visual field deficit. Peripheral vision on the left-hand side is gone'. In addition to these physical disabilities, Lisa explained that 'I have noticed two changes in my behaviour which have impacted the way that I work, and these are that I have lost most of my executive function… in terms of neurodivergent behaviours, the two things that I'm most aware of are, first of all, impulsivity, and secondly, disinhibition'.

DOI: 10.4324/9781032714745-14

At this point I should mention that Lisa's sense of positivity is so striking that, during her interview with me, she only once briefly reflected on the challenges that she faced during her recovery. Instead, she enthusiastically shared how she has returned to work, bringing her skills in marketing and business to a variety of projects, including a book on bringing hope to the lives of brain injury survivors. Rather than focusing on the challenges associated with her cognitive differences, she said, 'I can give you examples of how those two new behaviours can be helpful in business! I was always quite a confident person but I've kind of lost the inner critic, this is a voice that was sometimes stopping me from doing things because you're "not good enough" or "not qualified enough to comment or something". I don't have that any more; I just quite happily will pile in and share opinions and information and that's opened up a lot of opportunities'.

When we discussed the specific topic of 'neurodiversity', I noticed that Lisa preferred the term 'neurodivergence'. I asked what either of these terms meant to her, and she responded, 'I would say that I think that those behaviours are a diversion from normal expectations and from how I was behaving prior to the brain injury... that my family have noticed and remark on, changes in my behaviour that they always attribute to the fact that it's a 'stroke thing'; especially my impulsivity and my lack of inhibition'. When I asked whether her understanding of or identification with neurodivergence has contributed to her strong sense of hope and positivity, she said: 'No!' Instead, Lisa explained: 'I've got a very strong faith as well, that really underpins a lot of what I do'. She added, 'I have maybe had more difficulty communicating with [her children] as a result of my responses that can be quite unexpected at times, and I think that has made things a little bit more difficult'. When answering the question of 'what helps?', Lisa told me 'I actively take time to reflect on conversations and consider how things might have been able to be handled better. Because I live with neuro fatigue post stroke it means that to manage my fatigue I have to rest every day, 2–5pm, so that I have quiet time three hours every afternoon, where I can use that time to go over in my mind things that have happened and try and think of ways to improve them in future.' She also shared that she has 'done quite a lot of cognitive rehabilitation, to learn and develop strategies that I can apply to my life to genuinely make things easier for me and everyone who lives with me.'

It struck me, during our conversation, how thoughtful Lisa is with great awareness of her strengths and challenges. Lisa quickly jumps to what is hopeful in her life, even when reflecting on the times that can be hard. She said she does a lot now to support others with cognitive differences, for example, she sits on several committees for local hospitals, businesses, and has just been appointed spokesperson on behalf of patients who have had a stroke, contributing at board level in the National Health Service (NHS) in the UK, to help shape policies that affect stroke services. This work has introduced her to other people who would identify as 'neurodiverse' or 'neurodivergent'. In one of her groups, she explained: 'I have two members who live with a combination of brain injury and ADHD'. I asked what similarities Lisa has noticed: 'I'd say its planning and scheduling where we are very similar – we both need to have timetables and a fixed pattern to help us thrive'. She went on to note a key difference: 'I am able to implement self-control.' She clarified that her cognitive differences are directly related to the stroke she experienced, that is, they were not present in her earlier life, including childhood.

Lisa's positive account of how she now lives her life, is peppered with a sense of realism. She does not shy away from acknowledging her needs for support. At one point she shared with me that she 'literally couldn't get dressed in the morning' without the help of her carers. They drive her to events and meetings and help her participate in activities that are important to her. She tells me 'They're very good at making me stick to all my daily routines of exercise and rehab, which would be harder without somebody there, encouraging you to do it,'

Another key enabler for Lisa is technology. She described how online networking during COVID-19 made it easier for her to engage in social activities. 'I was quite happy to turn up and introduce myself on different platforms.' She caveats this with the potential dangers she is aware of when using technology: 'I am very, very aware that I am a vulnerable person, just in the last week my Facebook profile was hacked and cloned. I really don't like that there's someone out there pretending to be me, I find that really uncomfortable and it's making me more cautious about communicating anything online'.

I wouldn't want to leave this on a negative note: Lisa is the most positive force of energy you could imagine. My takeaway from the interview is that, while she is realistic about her needs and vulnerabilities, and she identifies as having neurodivergent differences, she does not let this interfere with living a fulfilling life. Her words resonate: 'I'm a lot more, kind of, brazen than I ever would have been before!'

Commentary

As is highlighted throughout this chapter, Lisa has a very positive outlook on her situation. Rather than dwelling on all the changes that have occurred since her stroke, she chooses to highlight the ways her life is still good despite these challenges. She acknowledges that she now has physical limitations that impede many aspects of her daily life, but she also is keen to point out that with the support she is given, she is able to tackle these head on.

Lisa states that she is most aware of the changes in her executive functioning, especially her impulsivity and disinhibition. She describes how this can cause barriers in her communication, in particular with her children, but also explains how at times this lack of inhibition can be useful, especially in a work context. She describes how before her stroke she would often hold back from giving her opinion due to fears of not being competent enough, however these fears no longer exist, so she is much more forthright in meetings.

When asked whether identifying as neurodivergent has helped her remain positive, Lisa is clear that this has not factored into her positivity. Nonetheless, she is able to see that many areas where she struggles are similar to areas where people with traditional diagnoses of neurodivergence (ADHD for example) also struggle. These areas include planning and scheduling. Lisa goes on to describe how having these difficulties as a direct consequence of her stroke makes it easier to implement strategies taught through cognitive rehabilitation than for those who have these difficulties from childhood.

Note

1 Lisa was interviewed by Sara Simblett.

15 Ashley's lived experience

Autism spectrum conditions (ASC), or just simply autism, is more commonly diagnosed in boys than girls. But why is this? Is there a genuine genetic difference? Or are we, as a society, more socialised into noticing the signs of autism that present in boys? And, if so, what is the consequence of this? There are many questions to be answered. But the fact of the matter is that sometimes autism is missed. People who have perhaps no understanding of why they perceive the world differently from others are encouraged to live lives that are in keeping with social norms. This can be a confusing and frustrating experience, and it is not uncommon for people in this situation to feel extremely stressed. The effects of this stress on the brain may include a disruption to its functioning, and lead to chronic problems with mood and cognition, on top of the experiences of undiagnosed autism, perhaps even exacerbating the misattribution of signs and symptoms to mental illness. Coming to terms with this is a process of grief and acceptance.

If I were to ask you, my neurotypical readers, what it is like to be neurotypical, how would you respond? Have you ever stopped to consider the question?

I have had to study this question all my life. Neurodivergence is all I know, and it is a fundamental part of who I am. My neurodivergence – my autism – shapes how I think, how I feel, how I see the world. I have had to think deeply about what it means to be neurodivergent in a world built by and for people who are neurotypical.

As a child, it was very clear that I was not like the other kids. Growing up in the 1990s, we did not have the language to discuss neurodiversity, let alone the criteria to diagnose it or the accommodations to support it. I was not formally diagnosed with anything until well into adulthood when the stress of my unmet needs led to serious physical health problems. In many cases, support needs and neurodevelopmental conditions are identified in children based on their ability to cope in school. In my case, I appeared quiet, well-behaved, intelligent, ambitious, and generally polite – the things teachers hope for in their students. What they couldn't see – and

DOI: 10.4324/9781032714745-15

what I didn't know how to express – is that I was also lost, lonely, and exhausted.

As I grew, I learned to watch others, figure out their patterns, and adapt myself to fit into their world. Every bit of this adaptation was done in a conscious, analytical way. In the autism community, we call this 'masking', and I was pretty good at it. But it came at a terrible cost. As I learned to adopt these patterns, I began to hide and even lose pieces of myself. I learned to ignore my own needs and to camouflage myself to meet others' expectations. The sensory environment and social structures of the class-room were overwhelming, and I was chronically burned out... as an 11-year-old. I clung to my schoolwork and books because that is where I knew I could succeed, and withdrew from human connections because that is where I knew I would fail. That pattern continued well into adulthood.

On paper, I was relatively successful. I excelled in school, got a master's degree and a PhD, and found jobs that I was passionate about. I maintained friendships and romantic relationships. However, it rarely occurred to me to ask my schools, workplaces, friends, or partners to adapt to me. To tone down the noise or the lights to ease my sensory overload. To meet in a quiet place in small groups, rather than at the bar or a party. To give me the space I needed to rest or respond. To be curious and open to my differences and quirks, rather than mocking them. Perhaps this was because I learned early on that such requests are often met with annoyance or worse. When I eventually started coming out as autistic and asking others for support, they often responded with some variation of 'I had no idea' or 'You don't look autistic', as if that was a positive thing. I repeatedly denied my own self and my own needs to fit in and 'succeed' in school, work, or social relationships.

Throughout this journey, my focus was on others. Though I was stressed most of the time, I wasn't really able to ask myself 'What am I actually feeling, and why?' I now know that this inability to identify one's own emotions, called alexithymia, is pretty common in autistic people. Instead, I focused on what everyone else felt, expressed, and wanted. I was unable to articulate my needs or boundaries, and sometimes I ended up in unsafe or traumatic situations as a result. Again, this penchant for ending up in trau-matic or dangerous situations is sadly common in autistic people, especially in autistic women. Many of us don't have the natural ability to recognise red flags, and our attempts to fit in can make us vulnerable to manipulation or abuse. I masked my way through countless cycles of burnout, depression, overdrive, and anxiety until I couldn't manage any more. I crashed. Hard.

Suddenly, but inevitably, I became disabled. For me, adopting the label of 'disabled' was empowering, because it allowed me (or forced me) to give myself space. It took several years of self-work and therapy to even under-stand what support I needed to ask for. The journey was not straightforward. The concept of a 'high-masking autistic woman' is still relatively new, and many healthcare professionals are not well versed in how neurodiversity presents across different genders and cultures. I was misdiagnosed or underdiagnosed several times. I had to actively pursue my autism diagnosis and the validation it eventually brought me. I was very privileged to have the educational background and financial means to do so.

A large part of my recovery was making my needs and differences known. Requesting or engaging in accommodations requires some level of vulnerability, especially when they make an 'invisible' disability suddenly visible to the people around you. For example, consider my personal nemesis: the grocery store. I need to eat to live, but on bad days its cacophony of colours, bright lights, people, and smells can leave me unable to walk or speak for hours. The deluge of input shuts down my brain. In the past, I would just avoid going. I would stop eating when the fridge emptied, which in turn worsened my sensory sensitivity. I have needed to adopt strategies to help me manage this essential activity in a healthy way. Noise-cancelling headphones work wonders, as does a supportive husband who is willing to take on more than his fair share of domestic responsibilities. When I need to shop, I always have a plan for getting through the store as quickly as humanly possible. But frequently, I also now need to publicly 'stim', which means making repetitive movements with my hands or feet to help me process and regulate the sensory overload. The first few times I allowed myself to stim publicly, I was desperately ashamed and self-conscious. However, it has become one of my most effective tools for managing my anxiety and overwhelm. This small thing – looking a little different and rocking back and forth a bit – has allowed me to engage more in life.

Visible social and physical quirks are typically all that others see of my neurodiversity. But often, those quirks are just reactions to living in an environment that doesn't fit my needs. The behaviours that others see are not what my autism 'is'. My experience is much deeper. So, what does it 'feel like' to be autistic? It differs for everyone, but for me, it boils down to two main things.

First, I feel and see everything at full volume. I do not have the subconscious filters that most neurotypical people seem to have that reduce life's background noise. I constantly hear the bathroom fan, the buzzing overhead lamp, the people around me breathing or shifting, and the cars outside. I can feel my clothes on my skin all day, every day, making me extremely sensitive to the textures and pressures they exert. I pick up all of the details from all of the work meetings. I am very aware of all the little movements that make up a person's body language, and the speech inflections which carry implied meaning. I am also intensely perceptive of other peoples' emotions, to the point where I have been described as 'empathic'. When I am in a communal setting, my perception is not limited to the people I am trying to engage with; I see and feel everyone in the room. This flies in the face of most autism stereotypes that I have encountered, but it is fairly common amongst those of us who identify as female or non-binary. Where I struggle is identifying the cause and formulating the 'appropriate' response to others' behaviours or emotions. It takes me time and analysis, and this can make me seem cold and aloof.

The second part of my experience is in the way I think. I think in connections, networks, and downstream effects. All of the data that my brain absorbs so openly is then connected to everything else. My experience of the world is shaped by the patterns that I so naturally see. This makes me very good at jobs that require you to find meaning and direction from noisy, chaotic information. It gives me a certain intuition and helps me predict others' behaviours or struggles fairly accurately. It also makes me tire easily – analysing that much information requires a lot of processing power. What I can't do very well is follow other people's abstractions or instructions. I need deep context and connections to understand what to do next. I am highly non-linear and can't retain recipes or step-by-step procedures to save my life. I need to understand the 'why' and I need to come to conclusions in my own way, otherwise, they may as well be in a different language. The heuristics and rules of thumb that seem to come so naturally to my neurotypical peers just aren't there in many cases. When I do have them, they have been carefully acquired rather than naturally developed.

My autism cannot be separated from me – it is just part of who I am, how I think, how I move, and ultimately how I interact with the world around me. Today, I know who I am and I love who I am, autism and all. I experience the world in a different way from most, and that is often a positive thing. I can naturally see patterns, make connections, and create in ways that my neurotypical peers cannot. I can feel intense joy from a special interest and intense connection with other people who are 'like me'. I have slowly but surely begun to unmask and allow myself to exist as I was born to. I sometimes struggle, but I have spent way too much energy hiding the differences and struggles from the people around me. From now on, I am unapologetically me.

Commentary

The final sentence of this chapter is what this entire book is about, and it is amazing that Ashley is not only able to feel it but also to write it down for all to read. If everyone who identifies as neurodivergent felt that they could be unapologetically themselves and society embraced this, then society would be moving in the right direction.

Before Ashley came to this point in her life however she had many obstacles to overcome. Much of these have been due to the perceptions of what an 'autistic person' is like – largely based on male autism, not taking into consideration the differences that women with autism face. Ashley describes how she excelled at school, partly because she delved into her schoolwork and partly because she learnt how to mask many of her differences. Although this resulted in her surpassing her education, Ashley highlights the impact that this masking had on her mental and physical health.

She describes how she felt exhausted and burnt out, as early as 11 years old, until she became disabled.

Once this inevitability had occurred, Ashley describes how she recovered from her disability by making her needs and differences known to others. This has not been an easy road, with people often commenting unhelpful statements like 'you don't look autistic'. Understanding her own autism and how this makes her different from neurotypicals seems to have been key to helping Ashley move towards recovery, not in that she is trying to be like the neurotypicals in her life but rather knowing what her strengths are so that she can build on these.

As Ashley points out, her autism is part of who she is, part of how she thinks, moves, and interacts with the world around her. Ashley has now become someone who can say that she loves herself as she is, autism and all, and that is quite an impressive statement!

16 Jack's lived experience

What if you've never been given a diagnosis of any particular thing but you know that you feel 'different'? Does this mean that the person is likely to be neurodivergent? Or is feeling 'different' just part of the normal experience of being human? Drawing on existentialist philosophy, people who feel different from others may experience existential loneliness, which can lead to existential dread. These feelings can contribute to feeling a disconnect from others, like others do not care about you, like one does not belong, or one is not fully understood. In this context, anxiety and depression are conceptualised as normal responses to feeling isolated or lonely. This book would not be complete without an exploration of how to interpret feelings of difference, from a lens that normalises human experience.

My name is Jack Versace. I am a 21-year-old master's student living in London. I study social psychology. Many of my academic interests lie in understanding the altered experience of self and the social world in people with severe forms of psychopathology, like schizophrenia. I would identify with the label neurodiverse; I believe that the unique ways I perceive the world and the experiences I have had have profoundly shaped who I am as a person. For a long time in my life, this was something I often struggled with and resisted. The questions these experiences raised, and the passions they stirred up in me, have spurred an interest in psychology and philosophy and pushed me to engage in prolonged personal reflection.

Therefore, I am glad to have been asked to contribute to this book. I see it as very important to try and demystify and humanise some of the more abstract ideas in neurodiversity, through a concrete discussion of lived experience. I also see it as an opportunity for self-empowerment, being able to discuss openly what was at a time a source of real insecurity and self-doubt.

Above all, I believe in the importance of the book's broader project: to expand the definition of neurodiversity and provide a platform for people to engage with this new definition and share their experiences. I am most

DOI: 10.4324/9781032714745-16

grateful for this opportunity to share my perspectives and reflect more deeply on what neurodiversity means to me.

I am going to proceed by discussing: (1) very briefly the editors' proposed redefinition of neurodiversity; (2) key experiences of mine that illustrate what neurodiversity means to me, and some lessons I learned from them; (3) the changes I would like to see at interpersonal and broader societal levels.

In the first chapter of this book, the editors argued that to be 'neurodiverse' is to differ from societal norms or expectations in terms of one's brain structure or function, as evident in observable differences in behaviour, emotion, or cognition. In their view, such neurodiversity could be permanent or temporary, and, crucially, does not necessarily imply damage or disability, but simply a different way of engaging with or experiencing the world. In principle, the many ways in which one could be neurodiverse vary as manifoldly as the brain itself does.

To begin with, there are several aspects of this definition that I deeply appreciate. For example, it emphasises non-pathologisation; I agree emphatically that neurodiversity is a celebration of natural variation, of unique ways of engaging with the world, with its own benefits and challenges, just as being neurotypical has its own benefits and challenges. Furthermore, I think it makes sense to expand neurodiversity to include all different ways the brain may function or be structured, not leaving it limited to just neurodevelopmental disorders. Indeed, I think the public often equates neurodiversity exclusively with the autism spectrum. Broadening the perception to embrace a more general and inclusive definition is crucial.

My view of neurodiversity, however, differs in one key respect. Specifically, I am concerned that overemphasis on depersonalised abstractions – such as differences in cognition, neural structure, or functioning – may obscure the real, lived experiences of individuals, or render them secondary. In that sense, I would emphasise neurodiversity primarily as being a matter of unique modes of 'being-in-the-world' (in the sense of the term used by Ludwig Binswanger, 1963). This term correctly highlights that neurodiversity is a matter of Being, not simply disjointed behaviours or emotions, but an entire existence. It also emphasises that such a Being is always embodied in the world; thinking simply in terms of abstractions may mislead us into thinking that we can separate our brain, mind, or personality, away from its world. Furthermore, the way in which someone is neurodiverse is not a fixed structure; rather, their life and their world are fundamentally open and dynamic.

Though in elemental ways they may differ in how they think or feel, their mode of being-in-the-world cannot be reduced to such differences; it has taken on its own form and meaning through a dynamic engagement with themselves, others, and the world. Therefore, I would say that, in understanding neurodiversity, we should privilege lived experience over neural differences. When I have thought about my neurodivergence in my life, it has always been at this more humanistic or existential level. Thinking about it in terms of cognition does not resonate with my day-to-day life, though certainly such differences exist, and understanding them constitutes a valid scientific enterprise.

Notably, despite prioritising neural differences in their definition, the editors did indicate how neurodivergent people 'interact with the world 'differently'. Further evidence of this more holistic stance is their decision to make the voices of people discussing their lived experiences central in the book. On that note, let us turn now to one such story.

At a young age, perhaps late in elementary school, I began to sense that something about me was different. It was unsettling because this was a difference I could scarcely understand or articulate. In a sense, vagueness was an essential part of it. And the anxiety that arose from such uncertainty caused me to live more or less falsely, catering to the perspectives and expectations of others. In hindsight, however, I must admit that I was highly successful in masking. I always had plenty of friends and generally fitted in, but this came at the cost of immense inner tension and confusion.

By late middle school and high school, the pressures to conform became overwhelming. I was exhausted by pretending to be interested in things I wasn't, to act in ways that felt unnatural to me, to present a face that was not my own. As my friends became sportier, so did I. Making this or that team became my new reward of social validation, the new end to which I feverishly strove. Many of the friendships felt somewhat fake and empty, but I felt trapped in the image I had made of myself, and the desire to be 'normal'. I suffered from intense social anxiety for a long time. On several occasions, I was supposed to go to large social events (especially high school football games), but the fear of having to talk with so many people, in combination with the general overstimulation involved, caused me to lie and make up excuses as to why I could not go. I hated feeling like I had to lie, but explaining everything to my friends seemed somehow more impossible. Predictably, I felt very alienated, alone, and powerless. My world felt closed off and predetermined; I was overwhelmed by a sense of injustice and frequently wondered *why I could not simply be like my friends.*

A related experience was the feeling that I needed to be special in some way. I put considerable pressure on myself in school. I wanted to get into an Ivy League university, to become very rich and successful. The less I felt accepted by others, the less I could accept myself, and the more I latched on to these external symbols of validation. In most respects, I felt I was living a life that was not my own, to ends alienated from my own feelings and ambitions: both in my interpersonal life and concerning a potential future career. I felt myself to be carrying an immense burden and a senseless one at that. I often wished that I could just be 'normal' and not feel all this resistance to myself. In retrospect, that resistance turned out to be my most healthy and vital instinct, being the source of my strength later when I began to try and live a more authentic life.

Perhaps worst of all, because I became so oriented to these false goals that I had become alienated from the things that I did enjoy. For example, the most characteristic aspect of my personality is my curiosity and open-ness. From when I was very little, I was always very nerdy. I became deeply engrossed in history and maths. Gradually as I grew older, I came to view this almost obsessive passion and curiosity as something strange and tried to distance myself from it. It was undermined further in two ways. On the one hand, I thought other people would regard it as weird, and I was sensitive about being ostracised for something I found personally meaningful. Sec-ondly, my interest in school gradually collapsed into my aforementioned struggle for high performance and symbols of validation. I felt very disillu-sioned, becoming gradually more separated from that primordial sense of childhood wonder and amazement. In its place, the world came to feel frozen, reified. Everything felt to me a matter of duty and conformity, not exploration and creation.

I felt puzzled and stuck, depressed. Life appeared to me as a passive thing that happened to me, not as my own world in which I could really be and actively transform. At the age of 16 a sudden and tragic event set me on the path of self-transformation. Completely unexpectedly, my grandfather passed away in the middle of the night. Being confronted so viscerally with death heightened the contradictions in my world. It brought them to the forefront and made me realise how untenable my situation was. My grand-father was, more than anyone I could point to in my life, the person with whom I could most simply and unconsciously be myself. He was just as nerdy as I was, just as passionate and curious. Losing him made me realise just how hollow and false many of my relationships were. It was a deeply sobering experience, rendering absurd many of the illusions I held and false compromises I made. For a time, I simply withdrew. I stopped talking to many of my then-friends, and because my previous desire for external sym-bols of success now appeared to me vacuous, I lost interest in trying in school. I felt alone and aimless; the shakiness of my once seemingly solid foundations put me face-to-face with an existential void. And yet, it was in this void, in the darkest and perhaps coldest period of my life that I found myself again, that I discovered a new courage and a new longing for life and the world.

After the funeral, I took some of my grandfather's books back home with me. Though he did not receive much of a formal education, he was an avid reader, a near-expert on Russian literature and history. For no particular reason, one day I began reading his copy of Dostoevsky's *Crime and Pun-ishment*. I was struck almost immediately by a feeling of recognition. Dos-toevsky was peering into my soul. Most importantly, for me at the time, was his ability to masterfully interweave feelings of despair, existential guilt, and despondency with hope and liberation. He captured the nausea and dizzi-ness of being divided against oneself and then opened up a path to self-reconciliation and redemption. Above all, I believe he sought to demon-strate that even from the depths of personal despair and alienation, one can emerge renewed, full of joy and at home in the world. In me, a new faith in life was budding, a realisation that I was not as hopelessly lost as I had thought. I reclaimed my passion. Indeed, it was the psychological depth of Dostoevsky that led me to read some of the works of psychoanalysts and existential psychologists, like Freud, Jung, Erich Fromm, and Viktor Frankl. Such reading sent me down the psychology rabbit hole, in large part determining the course my life has taken to this day. But above all, it helped me gain more self-insight and with that compassion, and acceptance. As Jung (2020) once said 'We cannot change anything until we accept it. Condemnation does not liberate, it oppresses.' I realised I had been fighting myself for years; taking the step back to just accept and affirm myself and this life was profoundly liberating.

And yet, I was to discover that existential self-analysis on its own was insufficient. The decisive change came when I moved to London and made friends with open-minded and interesting people. The combination of my heightened self-acceptance and the new environment where I could start fresh meant that I felt free to actualise myself in the world. Much of my prior anxiety came from trying to navigate and negotiate the social world. It was very empowering to realise that I was not as stuck as I thought I was. I had found the power within myself to change, to affirm everything, to have the 'courage to be' (Tillich, 1952/2014). Neither was the social world so deterministic, something that just *happened* to me. I had been thrown off by some of the close-mindedness where I lived and among some of my then-friends. Now I knew that I could shape my social environment, choose friends with whom I could find common ground, and be myself. I ceased to feel alone, and my social anxiety reduced dramatically. It is for these reasons that I would stress the personal and existential levels of neurodiversity as being key. Understanding and mutual recognition are only possible when you are willing to peer into the world of another person; not linger at surface level behaviour or emotions but look for the whole existence which gives them their meaning.

This existential and social struggle, and eventual triumph, is the main way I think about my experience of neurodiversity. That is not to say there are no other aspects, or other challenges I still face from time to time. Academically, I have to start my work very far in advance because I get overwhelmed by the stress of working at the last minute. Socially, I still have bouts of social anxiety here and there and have a generally pretty short social battery. But I've learned not to try to pathologise everything. I now know that if I am struggling with something, it is useless to feel guilty about it or convince myself I should feel otherwise. There is a deep peace in that acceptance. It has also helped me reinforce that I do not find my identity in my differences. I challenge the idea that the primary way for me to make sense of myself is to understand my distance from some supposed norm. Rather, I have found different outlets and ways to ground myself when I am feeling anxious or overwhelmed. Chief among them are meditation, reading, music, and talking with people who share my passions and interests. Each of these helps me, in one way or another, to transcend the day-to-day dramas and tasks, and feel more rooted in my fuller humanity.

Practical changes to institutions and accommodations that create more inclusive environments are very important. I do not wish in the least to undermine that. However, I do nevertheless fear that as there has been meaningful progress in this sphere in some ways (for example, many universities have shown deep commitment to accommodating students), there has been markedly less progress in actually understanding people. Indeed, schizophrenia, in particular, still faces a powerful stigma with many perceiving it as merely an aberration, something fundamentally incomprehensible.

As I hope to have pointed out in my story above, I think that each lived experience takes on a meaning of its own that cannot be reduced to cognitive differences. It was not practical accommodation that made a big difference in my life, but feeling understood and accepted, focusing on these issues from a cognitive or neuroscientific perspective is useful in many ways, especially as it concerns designing practical advice, interventions, and accommodations. But I do not think they have much to contribute to understanding people; as I have said above, this is certainly not how I think about my life. But that is why projects like this book that go out of their way to provide insight into the lives of real people and move beyond the abstractions, are so crucial. They help us transcend external forms of understanding rooted in difference, to approach the real person themselves, on their own terms. How this could be implemented into the broader social movement of neurodiversity, I am not sure. I believe a useful starting point would be integrating humanistic and existential psychology into a field dominated by neuroscience and cognitive science.

I would like to thank the reader for taking the time to read my chapter. I will conclude with two brief takeaway messages. Firstly, I want to reiterate that we should centre our attention on trying to help those who are the greatest victims of stigma and prejudice: people with schizophrenia. Secondly, we should be mindful that, although society and specific institutions, such as universities or employers, becoming more accommodating would be very important, we should also try to emphasise understanding people more. I fear that focusing just on practical solutions without building empathy and understanding may not lead to the change that is needed. Subtle forms of alienation could very well persist even in a world where all reasonable accommodations are implemented.

Commentary

Jack breaks his chapter into three parts: firstly he discusses what neurodiversity means to him, secondly, he describes his experiences with neurodiversity, and then thirdly he highlights what he would like to change, both interpersonally and within society.

Within the first section, Jack discusses how he believes that the term neurodiverse is in danger of depersonalising the lived experiences of people who identify as neurodivergent. He emphasises the neurodivergent experience as being more than just cognition or emotional responses, rather it is an entire existence.

Jack then goes on to describe his own experience with neurodiversity. He explains how he always felt that there was something different about him compared to others of his age, but that he masked much of this. He details how this would lead to him feeling exhausted but that he continued to try to 'fit in' with those around him. He explains how this led to feelings of loneliness and increased social anxiety. He describes how the death of his grandfather and the discovery of the author Dostoevsky helped him to find a new faith in life and realise that he was not lost as he had assumed. Through self-discovery and moving to a new location Jack was able to accept himself and actualise his position within his social circle.

Finally, Jack moves on to discuss what he would like to see change, both within himself and in society at large. He acknowledges that while there has been progress in society regarding trying to create more inclusive environments, he fears that there has been less progress in understanding the individual. He states that individual experiences cannot be reduced to cognitive differences and that what really made an improvement in his life was being understood and accepted.

References

Binswanger, L. (1963). *Being-in-the-world: Selected Papers of Ludwig Binswanger*. Basic Books.

Dostoevsky, F. (2001). *Crime and Punishment*. Signet Classics.

Jung, C. G. (2020). *Modern man in search of a soul*. Routledge.

Tillich, P. (1952/2014). *The Courage to Be* (3rd ed.) Yale University Press.

17 Simon Lees' lived experience

Epilepsy is a chronic neurological disorder that causes seizures, which are sudden bursts of electrical activity in the brain. Febrile convulsions, also known as febrile seizures, are seizures that occur in children with a high fever. They are the most common type of seizure in children. Life with epilepsy through childhood and into adulthood can feel very uncertain. It is often hard to associate seizure with a pattern of occurrence, so they often happen spontaneously and unexpectedly. This can be associated with a high level of risk. However, some people are able to find a medication that controls seizure occurrence. The consequences of and anxiety related to dangerous seizures does not leave the memory of people with epilepsy, but the greater level of control that medication can provide is often felt as a relief.

My experiences with epilepsy are somewhat unique. After experiencing 'febrile convulsions' at 18 months old, I lived with mild, undiagnosed epilepsy until I was officially diagnosed at the age of 26 in 1984. Looking back, I now recognise how these early experiences shaped my understanding of the complexity of the condition. In 1995, I sadly lost my elder brother from epilepsy, and then in 1999, nearly killed myself whilst driving home from work, having gone into status epilepticus due to forgetting my medication at lunchtime.

Epilepsy is just one example of a medical condition that most strident advocates claim to be a type of 'neurodiversity'. However, it is more widely accepted as just a 'neurological condition'. Many also classify epilepsy incorrectly as being a 'disease'. The more accurate description is that it's a symptom of an injury or malformation within the brain or nervous system. In my case, I have severe hippocampal sclerosis within the left temporal lobe, a malformation rather than an injury. Brain differences aren't deficits: more like variations that contribute to the diversity of the human experience. Genetic conditions co-occur with many overlapping symptoms. Also, the term 'neurodivergent' usually refers to an individual, whilst 'neurodiverse' is used to describe a group or a community.

DOI: 10.4324/9781032714745-17

In 1984 I'd experienced what was thought to be two tonic-clonic seizures. Electroencephalogram (EEG) tests diagnosed temporal lobe epilepsy during hyperventilation checks. I had experienced two or three, ten to 15 second long 'feelings' throughout my whole life. These are known to be focal aware seizures (simple partials), so mild I'd never shown physical outward signs to the onlooker or gone unconscious. In between clusters, some people think they are fully controlled so stop their medications, which can prove to be fatal.

Never realising I'd had a form of epilepsy, for many years I'd been able to drive, play the drums, play tennis, and run half marathons. I'd always thought these 'feelings' were a kind of 'déjà vu'. I would experience a feeling of disconnection from past experiences, but when I was close to realising their origin, my thinking processes would return to normal. Other symptoms included difficulty understanding the meaning of three or four lines of text whilst reading, and sometimes the inability to talk using correct words. I knew the words to use but wasn't 100% sure if they were coming out in the right order. Most of the time they were, but I wasn't exactly sure. Overall, I was hardly affected by the condition during childhood and into my mid-twenties, so I never felt the need to talk about it.

Ironically when experiencing my first ever tonic-conics at 26, from a cognitive point of view, I had started to get an idea just a few weeks earlier that I may have epilepsy, but like most people, my knowledge was sadly lacking. Most people just relate 'epilepsy' to falling to the ground then convulsing. These full-blown seizures involve a sudden lack of consciousness and then coming round slowly after a period of several minutes followed by confusion. Interestingly enough, one was when I was asleep, the other when totally awake. If a partner is next to you, a full witness account and explanation to a specialist can be extremely helpful.

Using a combination of anti-epileptic drugs, it took two years to achieve complete seizure freedom. I was therefore very fortunate to get to 26, not having to take numerous combinations of medications. It wasn't until 2001 when an MRI scan identified severe hippocampal sclerosis in my left temporal lobe, highlighted by the hippocampus being damaged, shrivelled, and smaller in size, compared to the one in my right temporal lobe.

My elder brother had been diagnosed in the early 1970s, but for a completely different reason. He had fallen down a flight of stairs following an accident at university. His death in 1995 had followed a second major seizure in one day due to having to change medications at that time. This change was not managed professionally at all. Only four years later in 1999, I came close to death myself driving for eight miles gradually losing consciousness behind the wheel. Forgetting medication at midday for the first time in 14 years I was very lucky to stop my car in time in a lay-by, before going into a 45-minute status epilepticus set of seizures. This seizure type is considered to be a medical emergency. Two very kind people noticing my erratic driving had pulled up behind me and phoned the police. Paramedics arrived quickly and together with staff at a local hospital managed to save my life.

An interesting aspect learnt since is that many people record all seizures in diaries. These records cannot be classed as accurate, as during sleep, I could easily have experienced many seizures. Many people tragically can die from SUDEP (Sudden Unexpected Death in Epilepsy) when these take place because a person's breathing and heart rate can be severely affected whilst asleep, especially if experiencing tonic-clonic seizures.

A huge problem with epilepsy compared to other medical conditions is its stigma. During the 1980s and 1990s I realised very few neurologists talked in detail about the condition, especially from a patient's point of view including how it can affect their quality of life. Due to the lack of time within consultations, all that was mentioned was medications and whether seizures had stopped, increased, or stayed the same in number per month. My GP was so helpful; he knew I wanted my driving licence back and encouraged me to join a national charity and learn about the condition myself. I even helped a government organisation in 2000 in London and to my astonishment found that out of 75 patients asked, not one single one had been helped or encouraged to learn about the condition themselves from a charity organisation, except for myself. My GP was also surprised as the cost was minimal.

A major aspect of its stigma involves people being frightened to talk about it to others due to what others may think about them. Job losses, driving, and not knowing when the next seizure will take place are major problems. Warnings on TV channels that 'flashing lights will appear during the next item report' is helpful, but very few realise that only 3–5% of people experience photosensitive epilepsy. The way perceptions have changed historically over centuries certainly hasn't helped. Medications also can be detrimental as side effects can be so different for each individual – a situation not helped by the existence of over 40 different types of seizure. I was lucky not to experience being frightened of many of the above, mainly because my brother was diagnosed before me. I'd personally seen little difference in him as we lived 60 miles apart. What did astound me though, was his reluctance to talk about it compared to myself, considering his normal confidence in life and how he went about it. My employers, unlike many, were actually very helpful on both occasions. I have little trouble mentioning my condition to anyone, but you need to be subtle and confident. I love watching their reactions when I mention I have epilepsy, as it teaches me so much about their perceptions of epilepsy.

Back in 2000, I studied the condition vocationally at university for three years and then became a trustee of one charity and a member of two others. During the last 25 years I have lectured and presented regularly at charity conferences and become part of many patient projects around the UK and Europe at universities and meetings for GPs/neurologists, nurses, pharmaceutical representatives, and lay audiences. Topics have included general epilepsy, epilepsy from a perspective, and SUDEP. My knowledge therefore has increased dramatically over the years, meeting other people with the condition, especially at conferences and making sure I attended as many lectures and face-to-face group sessions as possible.

Using epilepsy as an example within the subject of 'neurodiversity' will definitely help as people will talk about it a lot more, helping it be spoken about more frequently. Patients cannot criticise the general public if they don't take the initiative to be more open in the first place themselves. If they continue not to be, we will never get the help financially compared to virtually all other 'neurological' conditions. The differences are immense which is why this must be more accurately investigated and carried out professionally.

Commentary

Simon emphasises the problems that arise from underfunding in epilepsy research and the general understanding of the condition. As he points out, epilepsy is not a disease, rather it is a consequence of a malfunctioning brain. It can occur from birth due to genetic factors or complications that arise whilst the brain is still developing, or it can occur after brain insult or injury. Epilepsy, as in Simon's case, is often not a static condition, rather it can change over time. Simon describes how in his first few decades of life he experienced simple partial seizures but how these then developed into tonic-clonic seizures followed by status epilepticus.

Simon points out how epilepsy as a condition has been stigmatised throughout the ages. This is not necessarily aided by the fact that there are so many different types of epilepsy as well as many differing drugs used to treat the underlying cause, many of which have undesirable side effects. He highlights the fact that this stigma can often result in people not wanting to talk about their epilepsy and that employers are not always understanding.

Simon describes all of this very pragmatically but also stresses how having epilepsy himself propelled him into studying it at degree level. This has enabled him to go on to give lectures about epilepsy from not only a personal but also a professional standpoint; something very few other people are capable of doing. Finally, Simon acknowledges that seeing epilepsy under the neurodiverse umbrella can only be a positive thing, as then perhaps it will be discussed more openly.

18 Kate's lived experience

When medication fails to control seizure occurrence in epilepsy, the story is a bit different. In these cases, an alternative treatment is brain surgery. Brain surgery is a major procedure that involves removing or disconnecting parts of the brain to stop seizures. The most common type is resective surgery, where a surgeon removes the part of the brain that causes seizures. Although this procedure is as high as around 60–70% effective for reducing seizure occurrence in temporal lobe epilepsy, there can be other consequences. Side effects of brain surgery can lead to cognitive and emotional changes, similar to the effects of acquired brain injury (ABI).

I would not have always classed myself as neurodiverse; I had considered neurodiversity as more of a behavioural issue, more concerned with interaction and learning ability.

I have had epilepsy for as long as I can remember. As a child I suffered with partial seizures akin to daydreams and these developed into more severe delusions as I got older. When I reached my twenties, I was having weekly seizures which could not be controlled by medication. The different medications invariably made me drowsy and sometimes emotional. The seizures also restricted my independence due to being unable to drive and I would scream uncontrollably due to the delusion and intense sensation that someone was trying to kill me. This was embarrassing, especially when starting new jobs or classes where people did not understand what was happening.

By my late twenties, I also began having tonic-clonic seizures as well as weekly partial seizures. I eventually saw a private specialist in London who advised surgery to remove the part of my brain causing the seizures.

It was not until I underwent brain surgery in 2021 to remove my amygdala and part of my temporal lobe that I noted a change in my processing ability. Overall, though the surgery has been a success; I have now gone to having roughly only one seizure every six months.

I began noticing changes in my processing speed, roughly ten weeks after surgery, especially with word recall, short-term memory and processing of

DOI: 10.4324/9781032714745-18

visual information. It has been this experience which has led me to writing this book chapter.

I was warned before the surgery of various risks and the effect the procedure might have on my cognition, vision, and balance. However, I was also advised that the other side of my brain would compensate, and my brain would learn to 'reroute' itself in time.

I have worked as a solicitor since 2016, and my work requires me to read complex medical reports and statements, relaying this information succinctly. At the time of my operation, I was doing a lot of advocacy work and would be required to appear in court with very little preparation time and present ten to 20 cases a day, with court staff and other solicitors competing for my attention. This was always pressurised, but after the operation, it was clear the pressure was compounded.

Initially this led to me having to concede certain advocacy work; I was unable to fully process the visual information provided or present it effectively amidst interruptions and distractions. I felt worried that other people were always having to wait for me. I felt frustrated and embarrassed by this, and I thought people would think I was unprepared and mock me; when in fact I had been up late the night before in order to make sure I had taken in the information. I ended up burning myself out and when the chance came to move roles away from advocacy work in 2022, I seized the opportunity.

My mood ebbed and flowed throughout this initial process of getting to grips with the changes in my memory and cognition, as well as the trauma of the surgery. I struggled with low mood and feeling overly emotional from time to time. This was totally new ground for me, as I have always felt quite in control of my emotions and pragmatic.

Luckily, because of the surgery, I had access to rehabilitation and agreed to participate in a study looking at the impact on my cognitive function pre- and post-surgery. This meant I was able to have one to one sessions with a neuropsychologist. It was explained to me that my short-term memory had been affected by the surgery, but that the anxiety I was feeling was feeding into my struggling memory function and vice versa.

My neuropsychologist taught me techniques to overcome the difficulties I was facing. I always keep in mind now to remain goal focused and use strategies to manage the problems and thereby reduce the burden on my brain. Techniques I find useful include:

- Speaking/reading aloud when trying to learn and retain new information. If I have a long document to consider or proof read, I read aloud as I find I notice if I have missed information, as it does not scan. In addition, I make more use of tracked changes on Microsoft Word and notes on Adobe as I go along. This is so that I do not lose my train of thought by leaving the document to make separate notes. When able I will also print off documents I need to take in, so I can scribble notes in the margin and jot down key points as I read.

- Keeping a list of tasks and ticking off as I go along.
- Delegating tasks and organising by importance and making sure I stick to the set plan, even if I am pestered or pressured from others. For instance, I keep a rolling task list organised by urgency and even if I receive an email or document to read, I will not fully consider this until the task related to the document/email comes up on my list.
- Taking a breath and going through the alphabet if I struggle with word finding. In some cases, it has been finding a simple word such as 'fox' which has left me fumbling verbally.

I still have my moments where I panic and feel over emotional. I find if I try not to take myself too seriously this helps to counteract the feeling of being overwhelmed. To try to overcome these moments I have learnt to listen to my body and rest when I need to.

If society more generally understood neurodiversity, I believe there would be greater patience, tools, and adaptations for neurodiverse individuals. The ability is there, however the route taken is different.

I think that by expanding the definition to include anyone who at times experiences cognitive differences in memory, concentration, planning, problem solving, and/ or decision making, there would be a better understanding of neurodiversity and how broad-ranging it is. By doing this, it could help to negate any stigma the term has, and society would be more accepting. This is because perhaps more people may realise they display neurodiverse traits.

I think my experiences of neurodiversity have been relatively mild although they have been exacerbated by the preceding events and my work. I am now at a stage where my brain has 're-routed' itself, albeit this is not yet second nature and still takes effort, such that I get easily fatigued. Because of this, my emotions are more balanced, and I am almost back to how I was pre-surgery.

I understand now that neurodiversity simply means a different way of processing and expressing information, thoughts, and feelings. From my point of view, keeping this in mind does help to contextualise this new situation I find myself in. I found accessing professional rehabilitative care very helpful in getting to where I am now. In addition, by accepting that I have reached a certain point in my recovery, and by adapting my routines and processes day to day to suit my capacity level, the frustration I felt at myself has gone and I am no longer caught in a circular trap of frustration and struggling memory function.

Commentary

In Kate's chapter she describes how up until recently she would not have used the term neurodivergent to identify herself. She highlights the fact that she has had epilepsy since childhood and how this epilepsy has developed throughout her life. She goes on to explain how this was having a detrimental effect on her, so much so that she agreed to surgery to remove the part of her brain responsible for the seizures in 2021.

Kate describes the impact of this surgery on her everyday functioning. She has noticed changes in her word-finding, short-term memory, and processing of visual information. Although she had been warned that these effects would take place after the surgery, she had also been reassured that the other side of her brain would compensate, so many of these difficulties would be short lived. This is not what Kate has experienced: rather she found that these changes in her cognition have impacted her ability to carry out her job as a solicitor. This in turn resulted in worsening mood and heightened anxiety.

Kate details the neuropsychological input she received and how this was imperative for her understanding of how the anxiety she felt was impacting her memory and other cognitive abilities. Through these sessions she was taught strategies to overcome the difficulties she was now facing.

Kate goes on to say how she feels that if society in general understood the needs of those who identify as neurodivergent, there would be greater patience, tools, and adaptations made. She points out how broad-ranging neurodiversity is, especially if the diagnosis were to be expanded to include anyone who experiences cognitive differences in memory, concentration, planning, problem solving, or decision making.

19 V's lived experience

Although some people experience physical changes such as hemiparesis after a stroke, some do not. For some people changes are on the surface, invisible to others. While one might think that this puts these people in a strong position to adapt to the effects of a stroke, this can be a big misconception. The hidden cognitive changes may pose significant challenges in everyday life, especially when needing to explain to friends, family members, employers, etc. about one's specific needs. This challenge is further impacted if communication is affected by the stroke. While close others may notice differences in communication style, other more distant people, such as new colleagues, may misattribute certain characteristics and assume that a person has not only always been this specific way but is acting on purpose.

V and I met in central London, where I had booked a quiet space for us to meet and chat.[1] He greeted me with one of the biggest smiles and we commented on how lovely it was to finally meet in person. Nowhere in this interaction was I aware of anything that seemed uncomfortable or difficult in any way, but this is one of many constant reminders of how invisible disabilities can be. Once we had settled into a conversation, V told me 'I used to be a lot more talkative'. He went on to say, 'it sounds strange because I will still talk a lot in some settings, but not as much as I did before', and when he talks about 'before' he is referring to a stroke that happened seven years ago. He explained to me that the stroke affected the 'right frontal lobe' of his brain and that a simple example of his acquired differences is that 'before the stroke, when I got into an Uber [taxi], I would very easily start a certain conversation with the Uber driver and spend the whole 15–30 minutes talking the whole time and I realised that immediately after the stroke, that had changed completely; that as soon as I got in the car I didn't want to talk and so I didn't start a conversation and we might spend those 15–30 minutes in complete silence'. To V, it became apparent that there are some clear differences, to him, in the way he is now, as compared to before the stroke.

DOI: 10.4324/9781032714745-19

When I asked V why he was willing to have the conversation that we were having today, he replied 'I think the first thing that piqued my interest was the realisation that I could fit within your definition, in the sense of acquired neurodiversity, because I assume stroke survivors are one example of people that acquire neurodiversity in adulthood'. He went on to say, 'my understanding is that if you had an average approach that people would employ to sort out a problem, or to learn a new concept, and you see how some people fall outside of the normal way of approaching things, those outliers might well be defined as neurodiverse people'. We laughed at this point about how we had mutually piqued each other's interest, when I went on to ask him to explain how he has learnt to use very novel ways of approaching everyday tasks using artificial intelligence (AI) tools. V said, again with the biggest smile, 'Yeah, I'm not here to plug my book, by the way!', then more seriously, 'ever since ChatGPT came out as a tool in November 2022, that has provided a massive help at work, on various levels. I had a colleague ask me jokingly recently, whether I ever "didn't" use ChatGPT?!'. He went on to explain, 'I mostly use it at work to code, to write emails, where I need to soften the tone because often my first approach might sound more aggressive or direct. That's how I ended up writing a small manual about ChatGPT for stroke survivors'.

V never ceases to amaze me in our conversations, never more so when, despite the differences in communication that he has noticed, he still loves to do public speaking, an activity that we acknowledged together that many people dread! V expanded on this revelation by explaining 'I think that there are situations in which I feel slightly more in control, and speaking in public is one of them. I've had a few opportunities to speak in public after the stroke and I find that situation weirdly comforting. It's something that stayed with me from before and it's a situation that is different from being in a conversation with people where you can be interrupted'. He said, 'I've found that when it comes to small groups, they are slightly chaotic; I tend to shut down and stop talking and become very silent. That is very apparent in groups of friends but also in certain types of meetings at work'.

V's determination to overcome the things that now challenge him became very evident through our interview. He spoke about how he has sought advice from the human resources (HR) department at his place of work but felt that 'they have no idea of how to approach the situation, like they had no instruments or tools or knowledge to tackle the situation' and that he was left to do a lot of the research himself regarding his legal right to reasonable adjustments. He told me, 'I've recently become more involved with a disability group that is part of my company, and I'm trying to think more and more in terms of policies and things that can help colleagues with disabilities'. During his research he explained 'I found out that the UK is one of the few places in the world that do not have quotas on employees with disabilities. In Europe, for example, in Germany or Italy or France, you'd get a percentage of employees who would have a disability and that doesn't seem to be the case in the UK or the US'. He went on to clarify, 'I'm not sure quotas are the right answer, but it definitely feels like there are cases in which companies aren't to just ignore that there is a societal need to employ and retain people that have a disability'. It was also very clear to me that work formed a very important part of V's identity. I also noted at this point that V does identify as having a disability, which may not be the same for everyone who identifies as neurodiverse or neurodivergent. V expressed that in fact he is very open about identifying as a person with a disability. He said to me, 'I don't want to hide the fact that I am a stroke survivor – I spell it out in the summary of my CV'. With great sadness, V also explained that he feels there can, however, be negative consequences to being this open: 'How many times I've sent out that CV and someone who might have sounded interested at first, just disappeared'. He told me that he feels this could be a 'very common experience'. We reflected on how this could be viewed as discrimination, and his parting words on this subject were: 'It's hard to demonstrate'.

On top of the research that V has carried out into policies and legal requirements, in his 'down time' he told me that he can find himself relating to posts about neurodiversity on social media: 'I have found a couple of tips and tricks that we have tried to apply at home', 'we have moved all the fruits and vegetables to the side of the fridge, at least if we have a peach staring at us, we will remember to actually eat it!'. We spoke about attention being something else that he has noticed differences in since the stroke. Relating back to the challenge he faces in terms of speaking up and 'jumping in' when talking in small groups, he told me that difficulties with attention play a role in this when he needs to follow multiple people speaking. In terms of meetings at work, when in small groups, he told me: 'Transcriber tools and other AI' have been a 'game changer' to support challenges with concentration and restlessness; as a result of this, V finds himself needing to move around his environment at regular intervals. His office is set up in a way that really facilitates this but there are times, for example when he uses public transport or lifts, when he needs to make adjustments that he feels may seem unusual or even 'hilarious' to others: 'while I was hospitalised in Italy, we would get a lift to go up to the stroke unit, and this lift opened at every floor; because I felt the urge to get out they had to catch me and get me back in all the way up to the fifth floor!'. V also shared that he can now be quite 'impulsive', and he notices this most when he plays chess: 'I played almost immediately after the stroke just to check whether my brain was still working in that way. One of the very first things I did after the stroke was to ask for a chess board and play against my father but I found when doing timed matches that it's really hard for me to wait until the next move, so I'll play extremely fast for a player at my level!'.

Despite V's constant humour and cheerful attitude, I know that things have not always been easy for him. We have spoken on numerous occasions about how he has had to fight to be heard and truly listened to, which,

I must say, he is doing a good job at now (as the Chair of the Disability Affinity group at his workplace), but I am sure there have been days that have been tough with difficult decisions to make. I am very grateful for him sharing his views and being supportive of my ideas too!

Commentary

V's story is one of hope and perseverance. He sustained a right frontal lobe stroke which resulted in noticeable personality changes as well as acquired difficulties with attention, inhibition, and impulsivity. Nonetheless, V has not let any of these difficulties impede him from living life to the maximum. Rather, he has learnt strategies to compensate for these challenges, including the use of new technology such as ChatGPT. Artificial Intelligence is a fascinating novel development to be included in neurorehabilitation and it appears that V uses this to excellent effect.

V is now also a champion for others with a disability at his workplace. He does not shy away from acknowledging his disability, rather he uses it to help. He is very open about the fact that he is a stroke survivor and what that means to him. Although this is generally positive, he also recognises that sometimes this can cause hindrance, especially if the people he is talking to are not open to seeing the benefits of employing someone with a disability.

V's outlook and use of humour make this chapter a pleasure to read and the fact that he has a platform to be heard, in the form of his book but also as Chair of Disability Affinity, is truly impressive.

Note

1 V was interviewed by Sara Simblett.

20 Redefining accessibility

Accessibility can be defined as the quality of being easily understood or appreciated. It is not just the practical adaptations required in order for a person to be able to obtain or use something, it has a relational connotation that places emphasis on the role of others being available through a process of communication and compassion to realise the needs of another individual. This is not a single transactional event but an ongoing process that requires willingness to engage on both sides. This final chapter of the book is an overview of all the lived experiences that have been expressed and explored in each preceding chapter, beginning with a reflective note from an editor and leading to a discussion of the social dimension of accessibility. We leave the reader with practical suggestions and recommendations on how people who identify as neurodiverse or neurodivergent wish to be treated.

Note from an editor

Writing and editing this book challenged me deeply, in ways that I did not expect when I started out. When it came time to write my own lived experience chapter, I inadvertently adopted the lenses I use as an epidemiologist and designer. I thought about function, behaviours, symptoms, abilities and disabilities, accessibility, and environmental factors. I unconsciously referenced common diagnostic and classification frameworks (DSM-5, the Diagnostic and Statistical Manual; ICD-11, the International Classification of Diseases; and ICF, the International Classification of Functioning, Disability and Health) and other functional models of health or disability. As a researcher, these were tools I had been trained to rely on. I have been trained to ask, 'What can this person *do*, what does this person *want or need*, and how can I help this person *do more* of what they *want or need*?' This was the approach I started to take with my own story. Note the focus on behaviour and function.

As I struggled to write, I was struck by how inauthentic it felt. I knew from personal experience that, too often, this focus on function can be

DOI: 10.4324/9781032714745-20

detrimental. The times I have been the most 'functional' with the fewest visible symptoms were also the times I was the most anxious, depressed, and repressed. I have spent a great deal of time, energy, and grief on self-reflection to shed a function-first view of myself and my abilities. It has been the greatest struggle – and greatest relief – of my adult life to understand and accept myself as a neurodivergent person. Neurodivergence is an internal experience as much as or more than it is a functional or visible one. This internal experience is what I knew these models failed to describe.

In the following chapter, we discuss accessibility and inclusion as it relates to neurodiversity. Our primary goal is not necessarily to suggest ways that you, the reader, can increase the function and abilities of the neurodiverse people around you. Rather, it is to highlight some of the experiences and needs of neurodiverse people from our perspectives and in our voices, in the hopes that they better equip you – the researchers, policymakers, educators, employers, and the general public – to empathise with us and foster inclusivity through your choices and actions.

Redefining accessibility: the social dimension

When we first set out to create this book, our vision was to conclude by discussing inclusivity and accessibility in a fairly conventional way. Based on our (the editors) professional and personal experiences with neurodiversity, we planned to use the lived experience accounts to discuss accessibility in education, the workplace, and other common spaces in which people live their lives. Indeed, our lived experience contributors confirmed that this consideration is necessary for their well-being, independence, and productivity.

> In the right environment, my nervous system no longer needs 'managing' and keeping in a safe window of tolerance. In the right environment, I am not fighting to just exist in the space, but I can actually conserve that energy and channel it into something that is useful for others, that can create a positive impact on humanity.
>
> (Kanan Tekchandani)

However, as we wrote, read, and reflected, it became clear that there was no single set of recommendations which could make work, school, or the built environment more accessible for the broad spectrum of neurodiverse people. Everyone had their own experiences, needs, preferences, and coping strategies.

> No neurodivergent person has a one-size-fits-all way that they would want help or support. Generally, the easiest way is to ask.
>
> (D)

While there is no single experience of neurodiversity, all of our contributors discussed how neurodiversity affected the ways in which they engaged and connected with others. Accessibility and inclusion within social interactions was described most consistently and with the most unmet need. It became clear that, just like we had adopted a broader definition of neurodiversity, we also needed to explore a broader definition of accessibility.

> I do nevertheless fear that as there has been meaningful progress in [accommodations and inclusive environments] in some ways (for example, many universities have shown deep commitment to accommodating students), there has been markedly less progress in actually understanding people.
>
> (Jack)

Here we discuss accessibility within our interpersonal relationships and interactions, or *social* accessibility. For the sake of this discussion, we define 'social accessibility' as accessibility, inclusiveness, and accommodations in the way neurotypical and neurodivergent people perceive, judge, engage, adapt, and relate to each other. This type of accessibility applies to all aspects of life where people can engage with one another, including work, school, and the lives we lead within our built environments.

Empathy and understanding – the basis of social accessibility

Almost all contributors in this book described how aware they were – often from a young age – of being different. Sometimes differences were subtle, sometimes they were less so. Sometimes they were accompanied by physical differences or disabilities. These differences were difficult to describe to others, because the neurodivergent and neurotypical people do not share a common language or reference points to discuss internal experiences.

> It's like trying to paint a kaleidoscopic, constantly morphing shape with a paintbrush far too thick for any detail, for an audience that only sees in black and white.
>
> (Joe Kelly)

> After reflecting on all of this in my late 20s I realised, for the first time, I genuinely don't think, feel, or act like the majority of people I had met.
>
> (Andrew)

> I don't think anyone can understand it unless they have it. It's internal. Which I call a battle, and it's sometimes very tiring.
>
> (Katy)

Quite often, this disconnect resulted in shame, stigma, or ridicule by others. These differences, which had a neurological basis, often resulted in behaviours which were seen as 'not quite right' or socially unacceptable. When these behaviours and differences were met with negative responses, many contributors learned to mask their differences. They accommodated the neurotypical people around them in social environments, often without reciprocity.

> Yes, I am more able to conform to the social and work expectations of the neurotypical world, but this is a façade.
>
> (D)

> People have the expectation that you are 'normal,' that you do normal things, and when you step out of that line, there is no acceptance.
>
> (Katy)

> Often, my experiences aren't recognised until they impact someone else and are met with hostility rather than conversations about occupational adaptations that could be supportive.
>
> (Becki)

When those who were adept at masking revealed their neurodiversity to others, they were often met with surprise or even congratulated that they were able to hide their disabilities and differences so well. These responses, often from well-meaning neurotypical people, were invalidating. They made it clear that neurotypical people frequently did not see or understand the struggles neurodiverse people face on a daily basis.

> Such phrases are often well-meant, said to reassure us that we are 'normal,' but instead feeds the narrative all too common with invisible illnesses; if other people can't see it, is it actually happening?
>
> (Becki)

Nor did neurotypical people understand the depth of experiences afforded by neurodiversity. Many (though not all) neurodivergent people see their neurodivergences as parts of themselves. It affects the way they think and perceive the world. By masking their differences to the people around them, they deny the world of that richness and deny their peers from knowing them authentically.

> What is clear to me is that ADHD is not just something that I have. It is a fundamental part of my being.
>
> (Joe Kelly)

I agree emphatically that neurodiversity is a celebration of natural variation, of unique ways of engaging with the world, with its own benefits and challenges, just as being neurotypical has its own benefits and challenges.

(Jack)

Empathy starts with education and self-awareness

Most contributors pointed to a general lack of education when it came to neurodiversity; both in the general population, and in opportunities for neurodiverse people to educate themselves about their conditions. In many cases, lack of education perpetuated harmful stigma and stereotypes.

I think that MS is generally seen as a condition that causes physical disability, but the other effects are less well understood.

(Janice)

My only prior image of the condition was the classic movie caricature of a loud, disruptive little boy, which seemed very far from the introverted person that I have always been.

(Joe)

These stereotypes and the information accessible to neurodivergent people about their own conditions often reflected behaviours and differences that *neurotypical* people experienced, and did not speak to the realities of life as a neurodivergent or neurodiverse person. The effects experienced by our neurodivergent contributors were twofold: they continued to live in a world which was unaware of their experiences and needs, and they had limited access to resources which would help them understand themselves and their conditions.

Too much focus has been placed on understanding neurodivergence from a neurotypical perspective; with neurodivergent individuals being spoken for and spoken about in ways that have historically stripped our population of autonomy, and often dignity.

(D)

Visible social and physical quirks are typically all that others see of my neurodiversity. But often, those quirks are just reactions to living in an environment that doesn't fit my needs.

(Ashley)

Each of the contributors described their own journeys to self-awareness and self-acceptance. For some with acquired neurodiversity that included coming to terms with changes in their abilities and sometimes changes in their personal identities. For others with developmental neurodiversity, self-awareness and self-acceptance was a journey that took place over years and decades. Many described relief and peace when they finally got their diagnoses or met others with similar conditions. Across all stories, knowing oneself, accepting oneself, and living in an authentic way was both a deep desire and a source of pride.

> I also see it as an opportunity for self-empowerment, being able to discuss openly what was at a time a source of real insecurity and self-doubt.
>
> (Jack)

Many contributors described investing a great deal of effort learning about themselves. It is important to note that the contributors to this book are people who were fortunate enough to access care, support, and the space to build this self-awareness. Not all are so lucky. Many described becoming strong advocates for themselves when healthcare institutions failed them, accessing private or specialty care to get diagnoses or evaluations.

> I know there's nothing I can do for my own younger self now, but one thing I would really like to see is more effort being made in schools to help each child at least understand themselves better.
>
> (Joe Kelly)

> I was lucky to be in relatively small classes where work was tailored to my needs. I also saw the most wonderful private tutor who taught me all the 'mental tricks' or 'compensatory strategies' that I still use to this day.
>
> (Sara(h))

Neurodivergent people are forced to think deeply about how they think and behave in relation to neurotypical people. Allowing both children and adults of all neurotypes the time and resources to get to know themselves would benefit everyone, as would increasing access to stories and experiences of neurodiverse people.

> It helps us to remember that there is variety in the human experience, and the social norms we once believed in aren't always in the best interests of the majority.
>
> (Becki)

Social accommodations

In professional and educational environments, it is common to refer to and grant neurodiverse people 'reasonable accommodations' to help them do their jobs or achieve their goals. Accommodations are typically seen as a positive step that employers or educators can take to foster a more diverse and inclusive workforce or student body, respectively. In many cases, these accommodations are also legally protected rights.

In stark contrast, 'accommodations' in the context of relationships often carry a negative connotation, especially in relation to mental health and neurodivergence. In the scientific literature, this term often refers to managing destructive or harmful behaviour or enabling and 'giving in to' an undesired behaviour (Adams & Emerson, 2020; Brewer et al., 2018; Rusbult et al., 1991; Strauss, Hale, & Stobie, 2015). In the literature, and often socially, accommodation is framed as something that can undermine a positive change toward a desired set of behaviours. Indeed, some of our contributors described that some accommodations can be a double-edged sword, undermining their function instead of expanding it.

> When we are pitied or helped it has the potential to weaken us, to make us believe that we cannot do certain things – learned helplessness essentially. If we are not treated in this way, but collaborated with, dynamically, so that both parties learn to understand each other better, then we can be flexible and adaptable.
>
> (Kanan Tekchandani)

The problem with this view of social accommodations is when and how society defines behaviours as 'undesirable', and the level of agency that a neurodivergent person maintains to influence the accommodations which are offered to them. It is admirable to want to help neurodiverse people fully engage in society, with fewer symptoms and less social rejection. And to be clear, we are not advocating for tolerating behaviours which actively cause harm, either to a neurodivergent person or the people around them. However, the contributors to the lived experience chapters described a pervasive lack of empathy and accommodation when it came to social interactions and expectations.

In particular, contributors with developmental sources of neurodiversity described facing pressure – both overt and unspoken – to act and be a certain way. From an early age, this led to experiences of otherness, reduced confidence, and isolation. Many described knowing that they were different, and not quite knowing why. Many faced stigma and negative perceptions of their diagnoses or conditions.

> I was diagnosed as a child, so have always viewed myself in the context of being different from my peers; although even before my diagnosis I could feel a disconnect between myself and that of most kids around me.
>
> (D)

> It is actually quite debilitating, people will look at you and think, you are purposefully deciding not to do something… what they can't see is, that in the executive function of the brain there is a breakdown.
>
> (Patrick Litani)

Because they are not accommodated, neurodiverse people often learn to hide or 'mask' their differences and act more like the neurotypical people around them. Neurodiverse people reported accommodating their neurotypical peers, rather than their neurotypical peers accommodating them. This accommodation can take an emotional and physical toll on neurodiverse people who already struggle to navigate a world which was not built for them.

> Neurodivergent people do a great deal of work to understand and relate to neurotypicals and attempt to act 'normal' as to not make them feel uncomfortable. I only began to question in my adult life why the same is rarely afforded inversely.
>
> (D)

> I was exhausted by pretending to be interested in things I wasn't, to act in ways that felt unnatural to me, to present a face that was not my own.
>
> (Jack)

Neurodivergent traits have been long viewed as 'problem behaviour' by broader neurotypical society (Shiloh et al., 2023; Donnellan, Hill, & Leary, 2013). Some examples described in this book include repetitive self-stimulating movements (i.e., 'stimming'), decreased attention span, lack of focus (or hyperfocus), impulsiveness, bluntness, and differences in communication strategies. However, for neurodiverse people, these traits have a neurological basis and are central to their lived experiences. They are a result of a different way of being, rather than an active choice. Some of these traits may even play a role in the unique value neurodiverse people can bring to a society or play a role in coping strategies neurodiverse people use to navigate the world around them.

> I would like the person reading this to understand that there is more to neurodiversity than the 'naughty child in school'.
>
> (Andrew)

> I also now need to publicly 'stim', which means making repetitive movements with my hands or feet to help me process and regulate the sensory overload. The first few times I allowed myself to stim publicly, I was desperately ashamed and self-conscious. However, it has become one of my most effective tools for managing my anxiety and overwhelm.
>
> (Ashley)

It is important to understand that there are valid reasons for these behaviours. Encouraging neurodiverse people to mask their diversity can be traumatic and detrimental to neurodivergent people in the long term. Accommodation in our social relationships and environments does not need to be a negative thing. Rather, it can empower neurodiverse people to engage more authentically.

> If you know that somebody is not engaging, perhaps rather than making the automatic decision that the person is doing it on purpose, perhaps ask the question 'why?' What I'm advocating for is that there's more onus on society to be aware and supportive of those who have cognitive difficulties.
>
> (Patrick Litani)

While there is no one-size-fits-all approach to engaging with others (neurodiverse or otherwise), there are practices we can add to our daily lives which can make social spaces more accessible for neurodiverse people. We encourage people of all neurotypes to consider the following actions and accommodations:

Cultivate self-awareness. Almost counter-intuitively, neurotypical people can help their neurodivergent or neurodiverse peers by becoming more aware *of themselves*, and how that differs from other potential ways of being. As a neurotypical person, how do you think? How do you feel and express emotion? How do you prefer to communicate? Where are your blind spots? Be aware of your own frame of reference. Don't assume that everyone experiences the world the way you do – or even that most people experience the world the way you do.

Remember that neurodiversity is all around you. Neurodiversity is part of the human condition. Though it is quite common, it is often invisible. Neurodiverse or neurodivergent individuals who have experienced stigma are often highly skilled at hiding their differences from others. You may not know if and when the people around you have different neurotypes than you. Whenever possible, make accessibility your default.

Assess your communication style. Are you direct or indirect? How often do you rely on tone or hidden meaning? Are you comfortable with pauses while others formulate their response? How often do you step back and confirm whether you and your conversation partners are understanding each other accurately? Communication can be difficult when people experience the world in fundamentally different ways. Difficulties in spoken communication and personal interactions may be exacerbated by social anxiety, environmental overstimulation, or acquired injuries to language areas of the brain. Being flexible in your communication style may allow others to engage with you more easily.

Be a safe harbour. Being different makes neurodiverse people vulnerable and can threaten their physical and psychological safety. Too often, differences and vulnerability are met with hostility or derision, especially when the reason for those differences are not visible or well understood. Ask yourself when and how often others are vulnerable with you. Are you giving others the space they need to express themselves and their needs? Be the type of person who makes it safe for others to be vulnerable and authentic. Make it safe to ask for support.

Adapt when needed. Adopt 'reasonable accommodations' in your social interactions. Practice willingness to hear and respect neurodiverse peoples' needs. While not all neurodiverse people will need the same support or accommodations, most have a good idea of the types of support they need. Be curious and suspend stigma or judgement. Question your assumptions and what you think you know about a condition. Listen to them and learn from the people around you.

Ultimately, we all decide what is acceptable, what is 'normal', and how accessible we want to make the world. The onus is on each of us to empathise and connect with our fellow humans, and to see the world through different perspectives. It is in everyone's power to make the world a safer, more accommodating, more accessible place for neurodiverse people, whether the diversity is visible or not.

> 'Society' isn't an ethereal being that controls what is deemed acceptable and what isn't. It is the attitude of the collective. Without having space and platforms for minority groups to share and contribute, we can't call ourselves a diverse and inclusive culture.
>
> (Becki)

References

Adams, D. & Emerson, L. M. (2020). Family accommodation of anxiety in a community sample of children on the autism spectrum. *Journal of Anxiety Disorders*, 70, *102192*.

Brewer, G., Bennett, C., Davidson, L., Ireen, A., Phipps, A.J., Stewart-Wilkes, D., & Wilson, B. (2018). Dark triad traits and romantic relationship attachment, accommodation, and control. *Personality and Individual Differences*, 120, 202–208.

Donnellan, A. M., Hill, D. A., & Leary, M. R. (2013). Rethinking autism: implications of sensory and movement differences for understanding and support. *Frontiers in Integrative Neuroscience*, 6, 124.

Rusbult, C. E., Verette, J., Whitney, G. A., Slovik, L. F., & Lipkus, I. (1991). Accommodation processes in close relationships: Theory and preliminary empirical evidence. *Journal of Personality and Social Psychology*, 60(1), 53–78.

Shiloh, G., Gal, E., David, A., Kohn, E., Hazan, A., & Stolar, O. (2023). The Relations between Repetitive Behaviors and Family Accommodation among Children with Autism: A Mixed-Methods Study. *Children*, 10(4), 742.

Strauss, L., Hale, L., & Stobie, B. (2015). A meta-analytic review of the relationship between family accommodation and OCD symptom severity. *Journal of Anxiety Disorders*, 33, 95–102.

Index

Note: Locators in *italics* and **bold** refer figures and tables respectively

For Product Safety Concerns and Information please contact our EU
representative GPSR@taylorandfrancis.com
Taylor & Francis Verlag GmbH, Kaufingerstraße 24, 80331 München, Germany